AAPS Introductions in the Pharmaceutical Sciences

Volume 13

The *AAPS Introductions in the Pharmaceutical Sciences* book series is designed to support pharmaceutical scientists at the point of knowledge transition. Springer and the American Association of Pharmaceutical Scientists (AAPS) have partnered again to produce a second series that juxtaposes the *AAPS Advances in the Pharmaceutical Sciences* series. Whether shifting between positions, business models, research project objectives, or at a crossroad in professional development, scientists need to retool to meet the needs of the new scientific challenges ahead of them. These educational pivot points require the learner to develop new vocabulary in order to effectively communicate across disciplines, appreciate historical evolution within the knowledge area with the aim of appreciating the current limitations and potential for growth, learn new skills and evaluation metrics so that project planning and subsequent evolution are evidence based, as well as to simply "dust the rust off" content learned in previous educational or employment settings, or utilized during former scientific explorations. The *Introductions* book series will meet these needs and serve as a quick and easy-to-digest resource for contemporary science.

Anil B. Jindal
Editor

Pharmaceutical Process Engineering and Scale-up Principles

Springer

aaps®

Editor
Anil B. Jindal
Department of Pharmacy
Birla Institute of Technology
and Science, Pilani
Jhunjhunu, Rajasthan, India

ISSN 2522-834X ISSN 2522-8358 (electronic)
AAPS Introductions in the Pharmaceutical Sciences
ISBN 978-3-031-31379-0 ISBN 978-3-031-31380-6 (eBook)
https://doi.org/10.1007/978-3-031-31380-6

This Springer imprint is published by the registered company Springer Nature Switzerland AG
The registered company address is: Gewerbestrasse 11, 6330 Cham, Switzerland

This book is dedicated to my parents, Brij Bhushan Jindal and Kiran Jindal, my wife Renu, and my daughter Grisha for their immense love, constant encouragement, and unwavering support throughout the writing process.

Preface

In the pharmaceutical industry, the ability to effectively design, develop and scale up processes is critical for the successful production of safe and effective drug products. *Pharmaceutical Process Engineering and Scale-Up Principles* is a comprehensive guide that delves into the principles of process engineering and scale-up, which are essential for manufacturing pharmaceutical products. This book is intended to serve as a resource for undergraduate and graduate students in pharmaceutical science and technology, as well as pharmaceutical engineers and researchers in the field.

The book starts with an introduction to the properties of solids, which play a crucial role in various processes in pharmaceutical manufacturing. It then delves into the principles of process engineering and scale-up, including key unit operations of solid and liquid dosage forms. The book also covers advanced topics such as the manufacturing and scale-up of nanoformulations and the manufacture of biotechnology-derived pharmaceutical products.

Throughout the book, the authors have emphasized providing a balance between theoretical concepts and practical applications. The book is supplemented with various examples, case studies, and exercises, which will help the readers develop a deeper understanding of the subject. It is a well-structured and user-friendly guide with a clear and concise writing style that makes the material accessible to readers with varying levels of experience.

This book is a valuable resource for those who are seeking to build a career in the pharmaceutical industry or for those who are currently working in the field and wish to expand their knowledge of pharmaceutical process engineering and scale-up. It will also be a valuable resource for researchers, professors and scientists in the field of pharmaceutical science and technology.

This book will be an essential tool in understanding the intricacies of pharmaceutical process engineering and scale-up. I hope it will be a valuable resource for all its readers.

Pilani, Jhunjhunu, Rajasthan, India — Anil B. Jindal

Acknowledgements

I would like to express my sincere gratitude to all those who have contributed to the development of this book, *Pharmaceutical Process Engineering and Scale-Up Principles*.

This book would not have been possible without the invaluable contributions of leading experts from industry and academia. I am deeply grateful to all those who agreed to contribute to the book and have shared their knowledge and experience to make this book a reality. Their insights and expertise have greatly enriched the content of this book.

I would like to thank my peers in the pharmaceutical industry for sharing their expertise and insights. Their knowledge and experience have greatly enhanced the content of this book.

I would like to thank the AAPS Introduction to Pharmaceutical Sciences series editor and editorial board members for their invaluable suggestions during the proposal review process. I am also grateful to the publishing editor, Carolyn Spence, and the project coordinator, Gifty Priscilla, for their support during the manuscript submission and publication. Their contributions have been instrumental in shaping this book.

I extend my sincerest appreciation to my students for their invaluable assistance in formatting the manuscript to meet the publisher's requirements. Their tireless efforts and dedication have been instrumental in submitting the manuscript to the publisher.

Thank you all for making this book a reality.

Contents

Contributors

Lubna Ansari GOVT of NCT of Delhi, Delhi Institute of Pharmaceutical Sciences and Research (DIPSAR), Delhi Pharmaceutical Sciences and Research University (DPSRU), New Delhi, India

Sagar S. Bachhav Clinical Pharmacology and Pharmacometrics, AbbVie, Inc., North Chicago, IL, USA

Kailash Bansal Formulation & Development, Amneal Pharmaceuticals Pvt. Ltd., Ahmedabad, India

Derajram Benival National Institute of Pharmaceutical Education and Research-Ahmedabad (NIPER-A) An Institute of National Importance, Government of India, Department of Pharmaceuticals, Ministry of Chemicals and Fertilizers, Gandhinagar, Gujarat, India

K. Bharathi National Institute of Pharmaceutical Education and Research-Ahmedabad (NIPER-A) An Institute of National Importance, Government of India, Department of Pharmaceuticals, Ministry of Chemicals and Fertilizers, Gandhinagar, Gujarat, India

Meenakshi Kanwar Chauhan GOVT of NCT of Delhi, Delhi Institute of Pharmaceutical Sciences and Research (DIPSAR), Delhi Pharmaceutical Sciences and Research University (DPSRU), New Delhi, India

Clara Fernandes IPA-MSB's Bombay College of Pharmacy, Mumbai, Maharashtra, India

Dalapathi Gugulothu GOVT of NCT of Delhi, Delhi Institute of Pharmaceutical Sciences and Research (DIPSAR), Delhi Pharmaceutical Sciences and Research University (DPSRU), New Delhi, India

Manasi Jathar IPA-MSB's Bombay College of Pharmacy, Mumbai, Maharashtra, India

Anil B. Jindal Department of Pharmacy, Birla Institute of Technology and Science (BITS), Pilani, Jhunjhunu, Rajasthan, India

Vishvesh M. Joshi Alembic Labs LLC, West Caldwell, NJ, USA

Bhakti Kubal IPA-MSB's Bombay College of Pharmacy, Kalina, Santacruz (E), Mumbai, Maharashtra, India

Ketan Patel College of Pharmacy and Health Sciences, St. John's University, New York, NY, USA

Rikin Patel Intas Pharmaceuticals Ltd., Ahmedabad, Gujarat, India

Kedar S. Prayag Department of Pharmacy, Birla Institute of Technology and Science, Pilani, Jhunjhunu, Rajasthan, India

Dhwani Rana National Institute of Pharmaceutical Education and Research-Ahmedabad (NIPER-A) An Institute of National Importance, Government of India, Department of Pharmaceuticals, Ministry of Chemicals and Fertilizers, Gandhinagar, Gujarat, India

Alisha Sachdeva GOVT of NCT of Delhi, Delhi Institute of Pharmaceutical Sciences and Research (DIPSAR), Delhi Pharmaceutical Sciences and Research University (DPSRU), New Delhi, India

Sagar Salave National Institute of Pharmaceutical Education and Research-Ahmedabad (NIPER-A) An Institute of National Importance, Government of India, Department of Pharmaceuticals, Ministry of Chemicals and Fertilizers, Gandhinagar, Gujarat, India

Siddharth Salve National Institute of Pharmaceutical Education and Research-Ahmedabad (NIPER-A) An Institute of National Importance, Government of India, Department of Pharmaceuticals, Ministry of Chemicals and Fertilizers, Gandhinagar, Gujarat, India

Ila M. Sarode Department of Pharmacy, Birla Institute of Technology and Science (BITS), Pilani, Jhunjhunu, Rajasthan, India

Bhakti Kubal Shweta Sawant IPA-MSB's Bombay College of Pharmacy, Mumbai, Maharashtra, India

Amit Sharma National Institute of Pharmaceutical Education and Research-Ahmedabad (NIPER-A) An Institute of National Importance, Government of India, Department of Pharmaceuticals, Ministry of Chemicals and Fertilizers, Gandhinagar, Gujarat, India

Tanvi Warde IPA-MSB's Bombay College of Pharmacy, Mumbai, Maharashtra, India

Part I
General Considerations

Chapter 1
Properties of Solids

Ila M. Sarode and Anil B. Jindal

Contents

Abstract Properties of solids, including particle size, particle size distribution, particle shape, particle density, surface area, porosity, surface charge, flow, and cohesiveness, play a significant role during the processing of powder and determine the quality of the product. Understanding the role of properties of solids is extremely important in optimization of pharmaceutical processes. In this chapter, we have discussed about the properties in detail and their effect on the processes involved during the preparation of dosage forms, such as mixing, drying, compression, and coating.

Keywords Pharmaceutical processing · Solid dosage form · Properties of solids · Powder

---

I. M. Sarode · A. B. Jindal (✉)
Department of Pharmacy, Birla Institute of Technology and Science (BITS), Pilani, Jhunjhunu, Rajasthan, India
e-mail: anil.jindal@pilani.bits-pilani.ac.in

A. B. Jindal (ed.), *Pharmaceutical Process Engineering and Scale-up Principles*, AAPS Introductions in the Pharmaceutical Sciences 13,
https://doi.org/10.1007/978-3-031-31380-6_1

1.1 Introduction

The development and manufacturing of pharmaceutical solid dosage forms need extensive study of particle technology. Particle technology plays an important role in the formulation of oral tablets, capsules, powders, parenteral suspension, and topical aerosol. It involves the study of properties of solids and assessment of their performance individually and in bulk. The knowledge is then used for designing formulations, optimizing manufacturing processes, and producing quality products. The behaviour of the particles during manufacturing determines the safety and efficacy of the final dosage form (Hickey & Giovagnoli, 2018).

The physical properties of solids are assessed at all the three levels, including molecular, particulate, and bulk level. The molecular, particulate, and bulk levels deal with the properties of individual molecules, individual solid particles, and particulate species, respectively. These properties include particle size, particle size distribution, particle shape, particle density, surface area, porosity, surface charge, powder flow, and cohesiveness. Assessment of these properties is not sufficient, quantitatively measuring the properties is also necessary (Brittain et al., 1991).

The unit operations involved in the pharmaceutical industry include grinding, mixing, blending, milling, drying, compression, and coating. The properties of solids also play a vital role during storage, dosing, and discharge from the hopper (Leon Lachman et al., 1987). In this chapter, we have discussed in detail about the fundamental properties of solids and their influence on the processes involved in the manufacturing of solid dosage forms.

1.2 Particle Size and Particle Size Distribution

Particles are 3-dimensional entities and can be measured by a one-dimensional parameter (i.e., diameter if it is entirely spherical). However, most particles are not completely spherical; it has varying shapes and rough surfaces. Hence, equivalent spheres theory is the preferred method to determine particle size. According to the equivalent sphere theory, the diameter of the arbitrary particle is equivalent to the diameter of the sphere having the same characteristics, such as volume or surface area. However, these equivalent diameters vary from each other for a particle having an aspect ratio considerably larger than 1. Traditionally, the size of solids is defined in terms of mesh size. Table 1.1 states the characteristic term for the given size range (Allen, 1990; Wills & Finch, 2016).

Examples of equivalent diameters are as follows:

Equivalent volume diameter (D_v): It is defined as the diameter of the sphere which has volume equivalent to the particle.

Table 1.1 Size range of the characteristic size terms

Characteristic term	Size range (D_{90})
Nanoparticles	< 0.1 μm
Ultrafine	0.1–1 μm
Fine	1–10 μm
Medium	10–1000 μm
Coarse	1–10 mm
Very coarse	> 10 mm

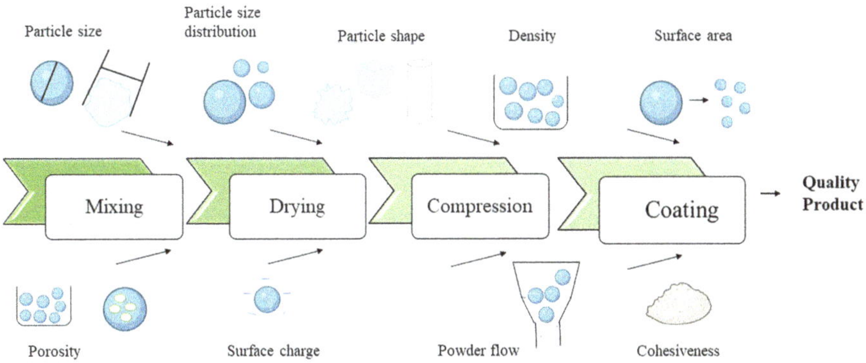

Fig. 1.1 Properties of solids influencing the manufacturing processes

Equivalent surface area diameter (D_s): It is defined as the diameter of the sphere which has surface area equivalent to the particle.

Stokes diameter (D_{Si}): It is defined as the diameter of the sphere which has a settling rate equivalent to the particle under the specified Stoke's law conditions.

Equivalent projected area diameter (D_A): It is defined as the diameter of the circle which has area equivalent to the particle's projection (Allen, 1990).

Based on the particle size, particles can be characterized as follows:

Particle size alone is not able to define the powder blend. It is important to measure the particle size distribution as well due to variations in particle sizes aroused during their production. Figure 1.1 describes the properties of solids influencing the manufacturing processes of powders. Narrow particle size distribution refers to a small variation in particle size. A mixture is said to be monodisperse when 90% of its particles lie within the median size range. Polydisperse powders are the ones in which the small and large particles have a large difference between their sizes. Table 1.2 states the terms used to define the powder blend (Allen, 1990; Seville & Wu, 2016).

The particle size can be measured by various techniques in the laboratory as described below:

1. Microscopy
2. Sieve analysis

Table 1.2 Particle size distribution range of the characteristic terms

Characteristic term	Particle size distribution (D_{90}/D_{10})
Mono-sized	<1.02
Ultra-narrow	1.02–1.05
Narrow	1.05–1.5
Medium	1.5–4
Broad	4–10
Very broad	>10

3. Sedimentation method
4. Electronic counting method
5. Reflectance FTIR technique
6. Dynamic light scattering

Optical microscopy measures particles in the range of 150–0.8 μm. Particles in the size range of 0.01–1000 μ and 0.001–5 μm are measured by electron microscopy and transmission electron microscopy, respectively. Sieving and sedimentation methods measure particle size in the range of 50–1500 μ and 1–200 μ, respectively. Electronic counting methods could measure particles from 0.4 to 1600 μ (Evans, 2003; Wills & Finch, 2016).

Particle size has a profound impact on the unit operations used for the manufacturing of solid dosage forms. It plays a significant role in ensuring uniform mixing and excellent flow properties of pharmaceutical solids. Ordered mixing is the preferred approach for obtaining homogeneous mixtures. Particles with 100 μm size enhance the ordering phenomenon, and 40 μm particles complete it (Hersey, 1975). A large difference in the size of the particles results in the segregation of the solids. Particles with large sizes flow easily as they have substantial amount of mass. Small-sized particles have more surface area and hence give rise to inter-particle friction and cohesive/adhesive forces between particles while mixing. This force causes the particles to agglomerate and prevents them from mixing uniformly. These cohesive forces create problems during the fluidization, granulation, and dispersion of the drug during manufacturing. For particles with a size less than 10 mm, van der Waals forces come into operation and reduce the flow property of the particles due to agglomeration. However, in some cases, the flow is improved as the agglomerated particle mimics the larger particle. Although the flow is improved, it results in non-uniform distribution of the small particles in the mixture. Further, small-sized particles make the mixture prone to oxidation, moisture absorption, or adsorption. Patel and Carstensen have reported that equal-sized particles result in uniform, and fast mixing with less segregation (Carstensen & Patel, 1977). If particles are of different sizes and one-sized particles do not fit in the spaces between the larger particles, then mixing is slow. If the small-sized particles can fit in the spaces between larger particles, then mixing is rapid, but segregation is also rapid. Uniform mixing of the particles ensures uniformity of dose. For example, coloured solids when mixed uniformly prevent the mottling of the tablets. The fine size of lubricants is necessary to coat the granular material and exert its effect. Ointments, creams, and pastes containing uniform-sized particles ensure greater stability, a good appearance, and a smooth finish (Musha, 2013; Nakamura et al., 2004). Particle size also affects other processing parameters such as drying and compression. Small-sized particles have a greater area for heat and mass transfer, thereby getting dried faster. A decrease in particle size leads to more bonding of particles and increased compactibility. Small-sized particles tend to resist compression. The optimum particle size is desirable for the manufacturing process to yield desired quality of the product (Cabiscol, 2020; Venables & Wells, 2001; Schulze, 2021; Wünsch et al., 2021).

1.3 Particle Shape

Particle shape refers to the orientation of the points lying on the border of the particle. It can be defined with the help of three scales, the macroscale, mesoscale, and microscale. The macroscale gives information regarding the three dimensions of the particles. It is the ratio of the dimensions. Mesoscale and microscale refer to the roundness, angularity, and surface rugosity, porosity, respectively. The most common methods to determine the particle shape are electron and optical microscopy based on the measurement of macroscale and mesoscale. Particle shapes can be columnar, plate, granular, dendritic, irregular, flake, fibrous, disk, lath, cubic, acicular, rod, sphere, etc., (Seville & Wu, 2016). Figure 1.2 represents different shapes of the particles. These shapes determine the properties of solids such as flow property and attrition and their performance during processing. Particle shape can be defined quantitatively by measuring elongation, aspect ratio, flakiness, and chunkiness. Elongation is the ratio of length to breadth, the aspect ratio is the ratio of maximum to minimum ferret diameter, flakiness is the ratio of breadth to thickness, and chunkiness is the reciprocal of the aspect ratio (Allen, 1981, 1990; Merkus, 2009; Schulze, 2021).

The roundness of the particle can be calculated by Eqs. 1.1 and 1.2 (Merkus, 2009).

$$\text{Roundness} = \left(\frac{D_A}{L}\right)^2 \tag{1.1}$$

$$\text{Roundness} = \left(\frac{P^2}{4\pi A}\right) \tag{1.2}$$

D_A is the diameter of the circle which has area equivalent to the particle's projection
L is the length of the particle
P is the perimeter of the particle.
A is the area of the particle.

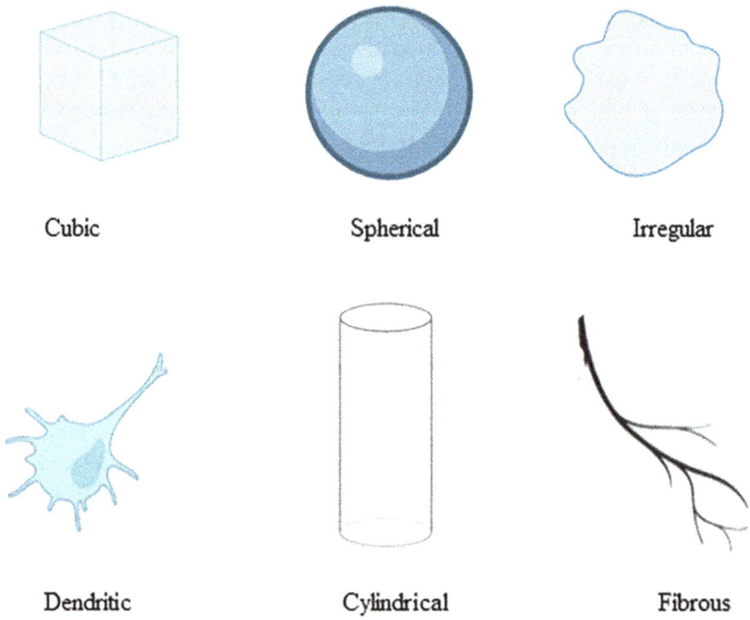

Fig. 1.2 Shapes of particles

The resemblance of the arbitrary particle to the sphere is described by the Waddell Sphericity (ψ) according to Eq. 1.3 (Merkus, 2009),

$$\Psi = \left(\frac{D_V}{D_s}\right)^2 \tag{1.3}$$

D_V is the diameter of the sphere which has volume equivalent to the particle.
D_S is the diameter of the sphere which has surface area equivalent to the particle.

The value of sphericity is less than one if the particle is not spherical. The inverse of sphericity is known as the coefficient of angularity or rugosity (Seville & Wu, 2016). Similar to particle size, the shape of the particles plays a vital role in the mixing, flowability, compression, and compactibility of powders. Spherical particles can be sheared easily as compared to angular particles. Angular particles have greater cohesive forces between them and pack tightly with each other. Elongated particles undergo interlocking of the microstructure. These factors lead to decreased flowability and non-uniform mixing. On the other hand, spherical particles roll over each other easily and ensure good flowability and uniform mixing. Cleary et al. studied the effect of spherical and super-quadric particles on mixing achieved by using a blade impeller. The super-quadric particles showed 14% and 4% lower radial and azimuthal mixing, respectively, and 3% higher vertical mixing than the spherical particles. The decrease in radial mixing was due to the enhanced resistance to shear by the elongated particles. Particle shape was found to have a distinct effect on different directions of mixing (Cleary et al., 2008). Particle shape has also been found to affect the compressibility and compactibility of the powder. It determines the type of interaction between the particles and the die wall. The particle positioning in a plane is determined by its shape and thereby determines the nature of bonding, such as solid bridges or interlocking. Particle roughness would lead to increased adherence of the particles to the die-wall causing the tablet defect known as sticking. A common problem seen in the pharmaceutical industry is the formation of needle-shaped particles upon crystallization. The needle-shaped particles show poor flowability and high compressibility. Poor flowability gives rise to non-uniformity in the release of contents from the hopper and die-filling. They generally form lumps upon drying. The particles are arranged uniformly in non-spherical particles and have a high surface area for contact with other particles. Hence, the compactibility of these particles is high. Particles should have optimum compactibility to exhibit tablet strength. Exceeding the limit results in compacts forming between the process leading to non-uniformity of dose. Particle shape also has a significant effect on the mechanical properties of powder and processing parameters of solid dosage forms (Merkus, 2009; Venables and Wells, 2001; Seville & Wu, 2016).

1.4 Particle Density

Particle density is one of the vital characteristics that influence the physical and mechanical properties of powders. It also determines the behaviour of the powder during processing (Merkus, 2009).

ISO 9276-1 (1) defines the fractional density and is denoted by the symbol q.

q_0 (D) is the fractional number density distribution
q_3 (D) is the fractional volume density distribution (Merkus, 2009).

Heywood described the different densities of the particles:

True density is defined as the ratio of mass and volume of the particle in which the open and closed pores are not included. Apparent density is defined as the ratio of the mass and volume including only the closed pores. The apparent density determines the quantity of the powder that will occupy the tablet die, capsule, or punch volume. Effective density is the ratio of the mass and volume including both the open and the closed pores. True density can be measured with the help of a helium pycnometer, or liquid displacement method for powder and bulk solids, respectively. Apparent density with the help of a mercury liquid pycnometer, and effective density by a liquid pycnometer having a low surface tension liquid. The liquid selected for the determination of true density should be such that the bulk solids get submerged in the liquid without dissolving in it. The liquid is placed in a measuring cylinder followed by the addition of a weighed amount of bulk solids. The initial and final volume after the addition of the bulk solid is measured and noted. The volume of the solids is equal to the difference in the initial and final volume readings

(Seville & Wu, 2016). The determination of the true density of powders using gas pycnometry involves the same principle. The liquid is replaced by an inert gas such as helium. Volume is determined by using the ideal gas law according to Eq. 1.4 (Seville & Wu, 2016).

$$PV = nRT \tag{1.4}$$

P = Pressure of gas
V = Volume of gas
n = Moles of gas
R = Ideal gas constant
T = Temperature of gas in kelvin

The gas pycnometer is made up of an expansion chamber and a measuring chamber and is linked by a valve. A weighed amount of solids are placed in the measuring chamber and air inside the chamber is drawn out. The chambers are closed by the valve and the gas generally helium is forced into the chamber and the pressure of the chamber is measured. The valve is opened and the gas diffuses in the expansion chamber till equilibrium is attained between the chambers. This method is usually important for the measurement of the density of dry fine powders (Seville & Wu, 2016; Leon Lachman et al., 1987). The method assumes a constant number of gas molecules which may not be true in all cases such as moist powder may vaporize some of its water content, increasing the pressure of the chamber. It results in erroneous estimation of the true density of the moist powder. In pharmaceutical processing, different solids are mixed and hence it is necessary to determine the effective true density of the powder. The mixing rule is used to define the density of the mixture (Seville & Wu, 2016). The density of the mixture is calculated using Eq. 1.5 (Seville & Wu, 2016):

$$\mathrm{p}M = \sum \mathrm{p}_{si}\xi_{\mathrm{i}} \tag{1.5}$$

p_{si} = Solid density of components
ξ_{i} = Volume fraction of components

For using the above equation, one must ensure that the components of the powder should be mixed with each other and do not segregate.

The bulk density is the ratio of mass and volume including the voids present between the particles. It is also known as packing density. Bulk density gives information regarding flowability, size storage spaces, volumetric feeders, and wall loading in the hopper. The packing of the powder is determined by the bulk density. It is measured by pouring the powders into a measuring cylinder or a cup; Amidon et al., 2017). Tapped density is the ratio of mass and volume calculated after tapping the powder. It is measured by pouring the powder into a measuring cylinder and tapping it. It takes into account the voids present after tapping. The volume obtained after tapping should be the lowest possible volume of the powder. Tapped density is greater than the bulk density and is not the powder's property but is dependent on tapping. The difference between bulk density and tapped density indicates the flow of particles with respect to each other. It gives information regarding the flowability and packing of the powder. Bulk density and tapped density are used to determine the voidage in the powder (Seville & Wu, 2016; Amidon et al., 2017). Voidage represents interparticle spaces with respect to the total volume. It is also known as a void fraction. On the other hand, porosity refers to the spaces within the particles. It is important to calculate the void fraction as it determines the tensile strength of the dosage form. The granules which have less bulk density and more intragranular porosity result in easy compaction and more tensile strength of the tablets (Espinal, 2012; Nimmo, 2013). Dense and hard would require more energy to compress them into a tablet. Compressibility index and Hausner's ratio are determined by bulk and tapped density, which is a measure of flow properties of the powder. Compressibility index measures the consolidation of the powder (Merkus, 2009; Seville & Wu, 2016; Nimmo, 2013).

Powders containing particles with different densities can cause several problems during the processing of pharmaceuticals. It leads to the enhancement of mixing time and probability of segregation. Particles having high density settle to the bottom due to gravitational forces and make the less dense particles remain at the top. Segregation may be observed on vibration because of the density difference in the particles. Particles with uniform and optimum densities are desired for obtaining quality products (Leon Lachman et al., 1987; Liebermann, 1990; Carstensen, 2000; Venables & Wells, 2001; Yamamoto et al., 2016).

1.5 Surface Area

The surface area of pharmaceutical solids plays a vital role in the determination of their physicochemical properties, thereby affecting the quality of the product. The mathematical formulas for the calculation of surface area such as $4\pi r^2$ for a sphere of radius r do not give the actual value as the surface has numerous imperfections. Microscopic examinations have revealed the presence of surface roughness because of the molecular and atomic orbitals present on the surface. Moreover, pores and voids also contribute significantly to surface imperfections. The actual surface area is greater than the theoretical surface area of the particle. The surface area of a powder is measured in terms of specific surface area which means the surface area of the particles present in 1 g of the sample (Lowell, 1931) as given in Eq. 1.6.

$$S_W = \left(\frac{\text{Surface area of particles}}{\text{Weight of particles}}\right) \tag{1.6}$$

Sw is the specific surface area

Pharmaceutical companies have specified that the lubricant, magnesium stearate should have a specific surface area of 25 m^2/g^{-1}. The particle size and particle shape affect the surface area of the powder. Small particles obtained by breaking a large particle possess greater surface area and a cubic particle and a spherical particle of the same weight and composition differ in the area by a factor of 2r/l (Lowell, 1931).

There are several methods for the determination of the surface area of a powder including particle size distribution, sedimentation, sieving, counting, gas adsorption, and air permeability. The gas adsorption method is most commonly used for the determination of surface area. Particle size distribution used for measuring the surface area is based on the assumption that the particles are spherical. The surface area of a spherical particle is calculated by $4\pi r^2$ and multiplied by the total number of particles. In the gas adsorption method, a gas such as nitrogen, argon, or krypton is allowed to get adsorbed on the surface of the solids under specific conditions. The adsorbed gas is used to calculate the surface area of the solid. Degassing is the first step of this method in which the surface impurities are eliminated including the adsorbed gases. It is required as the adsorption of gases on the surface would vary if contaminants are present. The presence of contaminants would cause a reduction in the actual value of the surface area causing errors in measurement. The solids are then cooled to cryogenic temperature and exposed to the gases at increasing pressures. The volume of the gas adsorbed at each point of increase in pressure is noted. The amount of gas required to form a monolayer over the solid gives the measure of the surface area of the solid. The air permeability method measures the rate of permeation of air in the bed of the powder which is dependent upon the surface area. It is advantageous over gas adsorption since it does not take into account the surface area of the cracks. The powder with measured porosity is filled in a cylindrical bed. Pressure is reduced along the cylindrical bed. The produced flow rate of air gives the surface area of the powder bed (Lowell, 1931; Akashkina & Ezerskii, 2000; Nimmo, 2013; Christian et al., 2018). Surface area is measured by the Kozeny-Carman equation as given in Eq. 1.7:

$$S_{ext} = \frac{14}{\Upsilon}\sqrt{\left(\frac{\delta_3}{1-\delta_2}\right)}\left(\frac{1}{\kappa \acute{\eta}}\right) \tag{1.7}$$

S_{ext} external specific surface area
Υ is density of sample
Δ is relative porosity
K is permeability of gas
Ή is viscosity of gas

Langmuir adsorption isotherm assumes that a cleansed solid is exposed to a gas. It assumes that the surface of the solid has N sites and gas molecules are adsorbed as a monolayer on the surface of the powder and energy E is the same for a particle at a site. The assumptions made in Langmuir adsorption do not hold true in many cases (Characteristics et al., 2021). Hence, the multi-layer adsorption theory was proposed by Braunauer et al. According to BET, the adsorption isotherms of all the solids follow one of the five patterns depicted in the figure. Type 1 adsorption isotherms are obtained when the adsorption of the gas takes place up to a few molecular surfaces. Microporous powders with a pore size of only a few molecular adsorbate diameters show type 1 isotherms. Non-porous powders having pores with a diameter greater than the micropores show type II adsorption isotherm. Pores having a radius of 50–1000 angstroms exhibit Type IV isotherms. Type V isotherms are seen in pores having radii same as that of Type IV isotherms and the adsorbate-adsorbent interaction same as that of Type III isotherms (Lowell, 1931; Characteristics et al., 2021).

The surface area has a profound impact on the processing parameters of pharmaceutical solids such as blending, purification, drying, tableting, compression, flowability, and coating (Leon Lachman et al., 1987). The more the surface area the greater the interparticle forces and the less uniform the mixing phenomenon. A greater surface area exposes more particles to the surroundings and hence more is the rate of drying. The powder that has a greater surface area would possess more compatibility due to the more surface area available for bonding. Surface area measurements determine the effectiveness of milling, especially for small particle-size solids. The surface area of the material to be granulated gives us the measure of the binder solution required to gain uniformity in the batches. More importance is given to surface area measurements since the advent of automated instruments for mixing and granulating. The vaginal tablets are designed in almond shape with a maximum surface area making them easy to disintegrate and disperse in the vaginal wall. Surface area is the vital characteristic of pharmaceutical solids necessary for efficacious delivery of the dosage form (Leon Lachman et al., 1987; Liebermann, 1990; Venables & Wells, 2001).

1.6 Porosity

Powder poured into a vessel produces a bed consisting of solid and void space. Porosity is defined as the ratio of the volume fraction of void space and the bulk volume of the sample. Porosity is the characteristic property of materials that arises during their preparation. Porosity comprises of micropores, mesopores, or macropores. Macropores have a diameter greater than 50 nm, and mesopores and micropores have diameters of 2–50 nm and less than 2 nm, respectively. Pore size and pore size distribution are important features of the powder which determine its behaviour and the quality of the final product. Pores are categorized based on their exposure to the surrounding fluid. Closed pores lie inside the solid and are not reachable to the surrounding fluid. It also affects macroscopic properties such as bulk density, mechanical strength, thermal conductivity, and elasticity. On the other hand, Open pores are accessible to the outer surface. These are divided into through pores and blind pores. Through pores pass through the solid and open at another end whereas blind pores are open at one end of the solid. These blind pores are responsible for the irregularity in the surface of the solid. The pores can be divided according to their shape including a funnel, cylindrical, slit, and ink bottle. Total porosity is equal to the addition of interparticulate porosity (porosity due to pores present between particles) and intraparticulate porosity (porosity due to pores present within particles), as given in Eqs. 1.8 and 1.9 (Espinal, 2012; Nimmo, 2013).

$$\text{Void volume } (V_v) = (\text{Bulk volume } (V_b) - \text{True volume } (V_t)) \tag{1.8}$$

$$\text{Porosity} = \left(\frac{\text{Void volume}}{\text{Bulk volume}}\right) \tag{1.9}$$

Several widely used approaches for the determination of porosity in industries are microscopy, gas sorption, X-ray and neutron scattering, and liquid intrusion (Espinal, 2012). Moreover, inverse size exclusion chromatography, positron annihilation spectroscopy, muon spin resonance, nuclear magnetic resonance, thermoporometry, and ultrasonic attenuation techniques are also some of the methods available for measuring the porosity of the sample. The method selected for measuring the porosity is dependent upon final use, availability of the equipment, pore-size range, powder attributes, and requirements. Microscopy is used for the measurement of pore size and geometry of mesopores by observing the cross-sections. Optical and electron microscopy are used based on the pore-size resolution required. Liquid intrusion is a method for measuring macropores and mesopores in the size range of 4 nm and 60 μm. The most commonly employed method is mercury porosimetry. Gas sorption is another method used for analysing micropores and mesopores. The porosity of the powder is calculated by methods based on physical adsorption. Type I isotherm symbolizes the presence of micropores and Type IV, V, and VI represents mesopores (Lowell, 1931; Espinal, 2012).

The porosity of the solids is important as it determines the type of packing the powder develops under storage, vibration, and when poured in a hopper, unloaded from drums (Leon Lachman et al., 1987). The directly compressible excipients have high intra-granular porosity and hence show high interparticle friction upon compression of the powder. The powder having high intragranular porosity is easily compressible and forms tablets of higher tensile strength. Porosity is a significant property of powder since it determines the deformation while compression, shelf-life, and moisture penetration (Leon Lachman et al., 1987; Liebermann, 1990).

1.7 Surface Charge

Processing of pharmaceutical solids in the industry involves sieving, triturating, or pouring the powders into the hopper. These processes cause the particles to come in contact with the surfaces of the equipment and get electrostatically charged by triboelectrification (Conway & Ghori, 2018). The produced charge may cause a change in the behaviour of the flow of the powder. A coincidental release of the build-up charge could cause fires, sparks, or explosions, hampering the safety of the process. Moreover, if the particles are poly-disperse, bi-polar charging occurs and could cause adhesion of the powder to the wall of the container, agglomeration, or segregation. The surface charge build-up on the powder is dependent upon the particle size. Small particles have a greater surface area, hence more contact with the wall, and more should be the build-up of charge. However, it is observed that the coarse particles gain more charge than the smaller particles. Saleh et al. explained that it occurs due to the deformation caused during the collision which enhances contact surface area proportional to the square of particle size. Mountain et al. explained it on the basis that fine particles have less inertia and propensity to strike the wall (Bailey, 1984; Zhang et al., 2015).

Initially, in the 1900s, it was believed that electrostatic charges developed on the powders were due to contact or friction between different substances. Further investigations indicated that electrostatic charges are induced due to contact and friction between particles of the same substance. In the following years, it was proved that contact and separation were responsible for the development of charge on the powder. Electrostatic charge is developed because of the transfer of electrons between the surfaces of the solids. Gold and Palermo studied the hopper flow electrostatics of acetaminophen formulations with the help of an ionostat (Jens Thurø Carstensen, 1980). Crystalline acetaminophen developed greater electrostatic charges as compared to the acetaminophen granules. Several excipients such as 2% lubricant and 0.5% water reduced the development of the hopper flow static charge. The electrostatic charges can be decreased by using an antistatic agent, humidity control, and alteration of the crystalline habit. Lachman and Lin have explained the equipment for calculating the inherent static charge and static electrification of powders. It includes humidifying chamber, recorder, and electrostatic tester. The electrostatic tester constitutes a power supply unit, a Faraday cage, and an electrostatic voltage sensing pistol. The powder sample is kept in the Faraday cage, and a decrease or accumulation is detected by the pistol and seen on the recorder (Reedyk & Perlman, 1968; Jens Thurø Carstensen, 1980).

The developed electrostatic forces have harmful effects on the mixing processes. Mixing of powders results in the production of static surface charges and powders form clumps after stirring for some time (Leon Lachman et al., 1987). Non-conductive substances that have the tendency to develop charge pose major static charge problems. In this case, enhancing surface conductivity is one of the remedial measures. The incorporation of surfactants, antistatic agents, glidants, and enhanced humidity can expend the build-up charge. Several pharmaceutical processes such as sizing, mixing, milling, and compression are responsible for the production of static charge. The build-up charge on the powder causes hindrances in the efficient performance of the processing equipment. These electrostatic forces do not affect the flow properties of granules prepared by slugging or wet granulation. These forces are negligible in comparison to the weight of the granules. Electrostatic forces play a predominant role in direct compression granulation. Zhang et al. stated that the introduction of fine particles leads to an increase in the rate of charge dissipation because of the enhanced area of contact of particles. The fine particles expend charge by acting as a lubricant around the coarse particles, and due to the high conductivity of these particles. Surface charge plays an important role in the smooth functioning of the processing equipment and delivering the product of desired quality (Jens Thurø Carstensen, 1980; Liebermann, 1990).

1.8 Powder Flow

The flow of pharmaceutical solids is important to the formulation scientist as it determines the discharge of granules from the hopper, uniformity of the dose, and compaction. Flowability measures the flow property of the powder. It is not the inherent characteristic of the powder rather it is dependent upon the process and physical properties of the solids such as size, shape, density, and moisture content. It is quantified using the angle of repose, Hausner ratio, Carr index, critical fill speed, and flow function (Wu et al., 2003; Krantz et al., 2009; Seville & Wu, 2016; Dubey, 2017). The angle of repose is the angle of the slope of the free surface of the bulk powder calculated under gravitational force. It gives us information regarding the flow behaviour of the powder. Figure 1.3 depicts the relation between the angle of repose and flow behaviour. The three types of angles of repose include the poured, drained, and dynamic angles of repose (Seville & Wu, 2016). The poured angle of repose is calculated by pouring the bulk solid from a specific height and measuring the slope angle. It finds application in selecting the appropriate design for the storage vessels. The drained angle of repose is the angle of the free surface of the bulk solid calculated after its discharge. It helps to determine the hopper design that will result in the excellent flow of bulk solids. The dynamic angle of repose is the maximum angle formed by bulk solids until they stay stable in rotating instruments such as drums. It is the angle measured just before the solids collapse during the motion. It is important to determine the dynamic angle of repose as it helps in designing the rotating equipments used in pharmaceutical industries and predicting the flow of

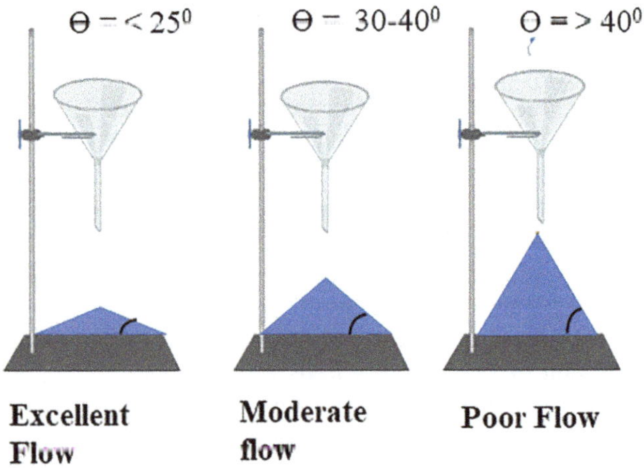

Fig. 1.3 Angle of repose depicting flowability of powder

solids in these types of equipments. A powder with a small angle of repose flows easily (Seville & Wu, 2016; Schulze, 2021). The angle of repose helps to determine the flow properties of the powder only if the cohesive forces between the particles are low or moderate. Powders with strong cohesive forces and a tendency to undergo segregation would result in erroneous results of estimation of flowability by the angle of repose. The flow of the powder from the hopper is classified into two regions, the central flowing region and the stagnant region. The stagnant region is present near the walls of the hopper and is dependent upon hopper angle and particle size. The stagnant region vanishes when the hopper angle exceeds 45°. The particles present at the bottom of the hopper flow freely. The particles at the centre have higher flow rate as compared to the particles present at the edges (Seville & Wu, 2016; Amidon et al., 2017). Hausner ratio and Carr index measure the flowability of bulk solids by bulk density and tapped density. Hausner ratio is tapped density divided by bulk density as given in Eq. 1.10. The Carr index is the difference in the tapped and bulk density divided by the tapped density of the powder as given in Eq. 1.11. The greater the Hausner ratio and Carr index, the poorer the flow of the powder, since cohesive powder voidage reduces on being tapped (Lau, 2001).

$$\text{Hausner ratio} = \left(\frac{p_T}{p_B}\right) \tag{1.10}$$

$$\text{Carr index} = \left(\frac{(p_T - p_B)}{p_T}\right) \tag{1.11}$$

$p_T =$ Tapped density
$p_B =$ Bulk density

Carr index, Hausner's ratio, and Jenike flow index give the measure of flowability of the powder. Carr index represents the compressibility of the powder. The more the compressibility index of the powder the easy it is to compress the bulk solid. Powder which has Hausner ratio of one is considered to have excellent flow and ensures reproducible die-filling. Jenike flow index describes how quickly the bulk solid might begin to flow (Seville & Wu, 2016; Stainforth & Berry, 1973). He quantitatively defined the beginning of flow by conducting confined and unconfined compression tests. The bulk solid is poured into a smooth container and subjected to consolidation by applying stress σ_1. The void spaces between the particles will be eliminated and the particles will reposition in the container. The particles of the powder exert stress sideways on the walls of the container which give rise to an inward stress σ_2 induced by the container walls. It can be explained as the stress experienced by the walls of the container because of the applied consolidation stress. Suppose we consider that the surface of the container is smooth, the frictional forces between the powder and the wall will be negligible and can be ignored. We again assume that the consolidated coherent powder can stand by itself without the help of the container. We remove the container and apply compressional force on the powder sample. A point will come where the powder will not be able to withstand

and will break and start to flow. The stress required to start the flow is known as the unconfined yield stress denoted by the symbol σ_c. The smaller the σ_c value the easier it will be to begin the flow of the powder at the given σ_1. Jenike flow index is calculated by dividing maximum consolidation stress by the unconfined yield stress. The index cannot be calculated for the small amount of samples as the coherent powder will not be formed at low amounts. The index is determined by dividing the maximum consolidation stress by the unconfined yield stress. The powder which has Jenike flow index greater than 10 is known to be free-flowing. The index value lying between 4 to 10, 2 to 4, 1 to 2, and less than 1 are known to be easy flowing, cohesive, very cohesive, and hardened, respectively (Seville & Wu, 2016; Stainforth & Berry, 1973).

Flowability is also measured with the help of flow meters most commonly for dynamic flow. The flow meters are named Hall and Flodex flow meters. The principle of the Hall flowmeter is based on measuring the time taken to dispense the bulk solids from a funnel of known orifice diameter. In hall flow meters 50 g of sample is allowed to pass through the funnel with an orifice diameter of 2.5 mm. The time taken for its flow is measured from which the mass flow rate is calculated. The Flodex flowmeter consists of a cylinder with a disk at the base which can be changed for the desired size of openings placed below the funnel. The bulk solids are passed through these disks until they get passed through the minimum-sized openings (Seville & Wu, 2016).

Compression of bulk solids requires that it should be free-flowing to fill the die cavity uniformly having the desired weight. Granulation of powders is required to improve the flow properties of the bulk solids. Free-flowing granules confirm uniform packing, superior packing density, and reproducible die-filling (Leon Lachman et al., 1987; Carstensen, 2000). Free-flowing powders avoid segregation and ensure uniform mixing. However, a free-flowing powder containing particles of different sizes and densities tends to segregate. The flow rate of the solids from the hopper should be high in order for the processing to be efficient and fast. Flowability is an important characteristic which determines the quality of the product (Liebermann, 1990; Carstensen, 2000).

1.9 Cohesiveness

Cohesiveness is the shear strength of powder in the absence of normal stress exerted on the plane of shear. Cohesive powders face several challenges in the pharmaceutical industry during their processing (Ghadiri et al., 2020). Cohesiveness arises in the granular bed due to electrostatic, van der Waals, and capillary forces. These cohesive forces in dry particles and wet particles are because of the van der Waals and capillary forces, respectively. The cohesive forces between the particles reduce the flowability of the powder. The behaviour of cohesive powder differs from that of dried powder which only has frictional interactions between the particles. The presence of cohesiveness in the powder is estimated by pouring the powder on a flat surface. Jenike described the poor flow of cohesive powders during storage and transport leads to bridging, changing flow rates, and channelling. The granular bond number quantitatively measures inter-particle cohesiveness. It is defined as the ratio of cohesive van der Waals force and the particle weight (Abdullah et al., 2010; Ghadiri et al., 2020).

The cohesiveness of a powder is measured by several parameters, such as Hausner's ratio, angle of repose, and direct shear test (Tomas, 2004). Initially, the annular shear cell was used for the determination of the cohesiveness of powder. However, owing to its challenges, the Jenike shear cell was introduced as the most reliable for the measurement of the cohesion of powders. It involves compaction and shear of powder on the application of a specific amount of load. Jenike described the flow factor as the ratio of consolidation stress to the unconfined yield strength. The flow factor of 1–2 and 2–4 indicates that the powder is highly cohesive and cohesive, respectively. The Johansen Indicizer is based on the principle of dissecting the powder after its compaction. The shear cell and tensile tester require more amount of powder and hence are not feasible (Shi et al., 2018). The angle of repose is calculated to study the cohesive behaviour of the powder. The cohesive force between the particles of the powder is observed when the powder is poured on a flat surface. The powder poured on the surface organizes into a conical pile. It would not occur in the absence of cohesive forces. The molecule present on the surface of the pile encounters two forces, gravitational and cohesive forces. The gravitational force acts tangentially and vertically downwards to the slant surface. The cohesive forces acting on the particle are perpendicular to the slant surface. The more cohesive forces in the powder, the more the angle of repose and the less the flowability of the powder. Brown and Richard described that fine cohesive powders possess an angle of repose greater than 40°. Hausner's ratio is used to determine the cohesiveness of powder. Dutta and Dullea explained that a less Hausner's ratio indicates a lesser degree of cohesion. Hausner's ratio of greater than 1.4 indicates that the powder is cohesive (Jens Thurø Carstensen, 1980; Shi et al., 2018).

The cohesiveness of the powder is vital as it affects the processing of pharmaceuticals such as mixing, compression, flowability, and coating. Cohesive powders have strong interparticle forces between them and form agglomerates, causing non-uniform mixing. The cohesive substances in the powder stick with each other and hence requires high shear forces to break the agglomerates for mixing to occur homogeneously (Qiao et al., 2012). High-speed granulators or fluid bed granulators are used to break these forces and mix cohesive powders. Mixing of mono-sized cohesive powders leads to segregation whereas segregation decreases in the case of bi-disperse powder. The flow of cohesive granules occurs in clusters because of the strong inter-particular forces. Moderate cohesivity is desirable in the powders and is important to assess it as it determines the caking, and flow behaviour of powder, mixing and segregation phenomenon, and the filling of cavities during compression (Shah et al., 2008; Jarray et al., 2019; Pasha et al., 2020; Zhu et al., 2020).

1.10 Conclusion

In recent years, the study of the influence of powder properties on the development, manufacturing, and quality of pharmaceutical products has undergone several advancements. Manufacturing of solid dosage forms from the bulk powder involves processes such as pouring of the powder in hopper, mixing, drying, compression, and coating. Mixing is mainly affected by the particle size, particle size distribution, particle shape, powder flow, and cohesiveness. Small-sized and irregularly shaped particles have greater interparticulate forces between them and hence form aggregates. If the particles present in the powder are polydispersed then it causes segregation of the powder. Drying of the wet mass is accelerated by converting it into granules. Granules have greater surface area as compared to the wet mass and the distance required to travel by the moisture to get dried is reduced. An excellent flow of the powder and optimum cohesiveness is necessary to ensure dose uniformity and acceptable behaviour during the manufacturing of the dosage form. Flow of the powders is also necessary to ensure that the die cavity is filled uniformly during compression of the granules. The fill space of the die cavity selected is dependent upon the density of the powder. More denser powder requires less fill space of die cavity as compared to less dense powder. Compression of the powder to form a tablet of desired strength requires optimum particle size, density, flow, cohesiveness, and porosity.

A lot of progress has been done in manufacturing solid dosage form and producing the product with desired quality. However, some unsolved issues relating the properties of powder to the processing of pharmaceutical dosage forms would improve the effectiveness of the processes and reduce the cost of production. Further, novel advancements in the study of the properties of solids and their influence on processing operations of dosage forms would give a new direction to the existing traditional approaches.

References

Akashkina, L. V., & Ezerskii, M. L. (2000). Determination of the surface area of powdered drugs by the air permeability technique. Part 1: PSKH instrument. *Pharmaceutical Chemistry Journal, 34*(6), 324–326. https://doi.org/10.1007/BF02524416

Allen, T. (1981). Particle size, shape and distribution, Particle Size Measurement. Springer, Boston, MA, pp. 103–164. https://doi.org/10.1007/978-1-4899-3063-7_4

Allen, T. (1990). Particle size measurement. https://doi.org/10.1007/978-94-009-0417-0

Amidon, G. E., Meyer, P. J., & Mudie, D. M. (2017). *Particle, powder, and compact characterization*. Developing Solid Oral Dosage Forms: Pharmaceutical Theory and Practice: Second Edition, Elsevier Inc., 271–293. https://doi.org/10.1016/B978-0-12-802447-8.00010-8

Bailey, A. G. (1984). Electrostatic phenomena during powder handling. *Powder Technology, 37*(1), 71–85. https://doi.org/10.1016/0032-5910(84)80007-8

Brittain, H. G., et al. (1991). Physical characterization of pharmaceutical solids. *Pharmaceutical Research: An Official Journal of the American Association of Pharmaceutical Scientists*, 963–973. https://doi.org/10.1023/A:1015888520352

Cabiscol, R., Shi, H., Wünsch, I., Magnanimo, V., Finke, J. H., Luding, S., & Kwade, A. (2020). Effect of particle size on powder compaction and tablet strength using limestone. *Advanced Powder Technology, 31*(3), 1280–1289.

Carstensen, J. T. (2000). Advanced pharmaceutical solids. *Advanced Pharmaceutical Solids, 110*, 1–510. https://doi.org/10.1201/b16941

Carstensen, J. T., & Patel, M. R. (1977). Blending of irregularly shaped particles. *Powder Technology, 17*(3), 273–282. https://doi.org/10.1016/0032-5910(77)80031-4

Characteristics, S. et al. (2021). '= P – M J', (2011), pp. 2020–2022.

Christian, R., et al. (2018). Development of biodegradable injectable in situ forming implants for sustained release of Lornoxicam. *Current Drug Delivery, 16*(1), 66–78. https://doi.org/10.2174/1567201815666180927155710

Cleary, P. W., & Sinnott, M. D. (2008). Assessing mixing characteristics of particle-mixing and granulation devices. Particuology, *6*(6), 419–444.

Conway, B. R., & Ghori, M. U. (2018). Triboelectrification of pharmaceutical powders: A critical review. *British Journal of Pharmacy, 3*(1). https://doi.org/10.5920/BJPHARM.2018.08

Dubey, A. (2017). Powder flow and blending. *Predictive Modeling of Pharmaceutical Unit Operations*, 39–69. https://doi.org/10.1016/B978-0-08-100154-7.00003-X

Espinal, L. (2012). Porosity and its measurement. *Characterization of Materials*. https://doi.org/10.1002/0471266965.com129

Evans, R. J. (2003). *Microscopy* (pp. 765–775). Encyclopedia of Physical Science and Technology. https://doi.org/10.1016/B0-12-227410-5/00444-0

Ghadiri, M., et al. (2020). Cohesive powder flow: Trends and challenges in characterisation and analysis. *KONA Powder and Particle Journal, 37*(October), 3–18. https://doi.org/10.14356/kona.2020018

Hersey, J. A. (1975). Ordered mixing: A new concept in powder mixing practice. *Powder Technology, 11*(1), 41–44. https://doi.org/10.1016/0032-5910(75)80021-0

Hickey, A. J., & Giovagnoli, S. (2018). Pharmaceutical powder and particles. https://doi.org/10.1007/978-3-319-91220-2

International, P. and Publications, S. (2010). Cohesiveness and flowability properties of silica gel powder E. C. Abdullah, A. M. Salam and A. R. *Aziz Department of Chemical Engineering, Faculty of Engineering, University of Malaya, 50603 Kuala Lumpur, 1*(1), 16–21.

Jarray, A., et al. (2019). Cohesion-driven mixing and segregation of dry granular media. *Scientific Reports, 9*(1), 1–12. https://doi.org/10.1038/s41598-019-49451-z

Jens Thurø Carstensen. (1980). Solid pharmaceutics: Mechanical properties and rate phenomena

Krantz, M., Zhang, H., & Zhu, J. (2009). Characterization of powder flow: Static and dynamic testing. *Powder Technology, 194*(3), 239–245.

Lau, E. (2001). Preformulation studies. *Separation Science and Technology, 3*(C), 173–233. https://doi.org/10.1016/S0149-6395(01)80007-6

Leon Lachman, et al. (1987). *The theory and practice of industrial pharmacy*. Varghese Publishing House.

Liebermann. (1990). *Pharmaceutical dosage forms: Tablets*. Informa Healthcare.

Lowell. (1931). *Powder surface area and porosity*. Powder Technology Series.

Measurements, P. S. (1975). Particle size measurement. *Dechema Monograph, 43*, 333. https://doi.org/10.4011/shikizai1937.43.333

Merkus, H. G. (2009). Particle Size Measurements, Fundamentals, Practice, Quality. Springer Science & Business Media, 17. https://doi.org/10.1007/978-1-4020-9016-5

Musha, H., et al. (2013). Effects of size and density differences on mixing of binary mixtures of particles, Powders and Grains. https://doi.org/10.1063/1.4812037

Nakamura, H., et al. (2004). EŠect of particle size on mixing degree in dispensation. *The Pharmaceutical Society of Japan-Regular Articles, 124*(3), 135–139.

Nimmo, J. R. (2013). Porosity and pore size distribution. *Reference Module in Earth Systems and Environmental Sciences*. https://doi.org/10.1016/B978-0-12-409548-9.05265-9

Pasha, M., et al. (2020). Prediction of flowability of cohesive powder mixtures at high strain rate conditions by discrete element method. *Powder Technology, 372*, 59–67. https://doi.org/10.1016/j.powtec.2020.05.110

Qiao, Z., et al. (2012). PVAm–PIP/PS composite membrane with high performance for CO_2/N_2 separation. *AIChE Journal, 59*(4), 215–228. https://doi.org/10.1002/aic

Reedyk, C. W., & Perlman, M. M. (1968). The measurement of surface charge. *Journal of The Electrochemical Society, 115*(1), 49. https://doi.org/10.1149/1.2411001

Schulze, D. (2021). *Powders and bulk solids*. Powders and Bulk Solids. https://doi.org/10.1007/978-3-030-76720-4

Sci-Hub. (n.d.-a). Assessing mixing characteristics of particle-mixing and granulation devices. *Particuology, 6*(6), 419–444. https://doi.org/10.1016/j.partic.2008.07.014

Sci-Hub. (n.d.-b). Characterization of powder flow: Static and dynamic testing. *Powder Technology, 194*(3), 239–245. https://doi.org/10.1016/j.powtec.2009.05.001

Sci-Hub. (n.d.-c). The flow of powder into simple and stepped dies. *Powder Technology, 134*(1–2), 24–39. https://doi.org/10.1016/S0032-5910(03)00130-X

Seville, J. P., & Wu, C. Y. (2016). *Particle Technology and Engineering:* An Engineer's Guide to Particles and Powders: Fundamentals and Computational Approaches. 1st edn. Butterworth-Heinemann. https://doi.org/10.1016/B978-0-08-098337-0.00001-1

Shah, R. B., Tawakkul, M. A., & Khan, M. A. (2008). Comparative evaluation of flow for pharmaceutical powders and granules. *AAPS Pharm Sci Tech, 9*(1), 250. https://doi.org/10.1208/S12249-008-9046-8

Shi, H., et al. (2018). Effect of particle size and cohesion on powder yielding and flow. *KONA Powder and Particle Journal, 2018*(35), 226–250. https://doi.org/10.14356/kona.2018014

Stainforth, P. T., & Berry, R. E. R. (1973). A general flowability index for powders. *Powder Technology, 8*(5–6), 243–251. https://doi.org/10.1016/0032-5910(73)80089-0

Tomas, J. (2004). Product design of cohesive powders – Mechanical properties, compression and flow behavior. *Chemical Engineering and Technology, 27*(6), 605–618. https://doi.org/10.1002/ceat.200406134

Venables, H. J., & Wells, J. I. (2001). Powder mixing. *Drug Development and Industrial Pharmacy, 27*(7), 599–612. https://doi.org/10.1081/DDC-100107316

Wills, B. A., & Finch, J. A. (2016). *Particle size analysis* (pp. 91–107). Wills' Mineral Processing Technology. https://doi.org/10.1016/B978-0-08-097053-0.00004-2

Wu, C. Y., Dihoru, L., & Cocks, A. C. (2003). The flow of powder into simple and stepped dies. *Powder Technology, 134*(1–2), 24–39.

Wünsch, I., et al. (2021). The influence of particle size on the application of compression and compaction models for tableting. *International Journal of Pharmaceutics, 599*(February), 120424. https://doi.org/10.1016/j.ijpharm.2021.120424

Yamamoto, M., Ishihara, S., & Kano, J. (2016). Evaluation of particle density effect for mixing behavior in a rotating drum mixer by DEM simulation. *Advanced Powder Technology, 27*(3), 864–870. https://doi.org/10.1016/J.APT.2015.12.013

Zhang, L., Bi, X., & Grace, J. R. (2015). Measurements of electrostatic charging of powder mixtures in a free-fall test device. *Procedia Engineering, 102*, 295–304. https://doi.org/10.1016/j.proeng.2015.01.146

Zhu, L., et al. (2020). Hopper discharge of cohesive powders using pulsated airflow. *AIChE Journal, 66*(7), 16240. https://doi.org/10.1002/aic.16240

Part II
Solid Dosage Forms: Pharmaceutical Process Engineering and Scale Up

Chapter 2
Mixing

Kailash Bansal and Anil B. Jindal

Contents

Abstract Mixing is an important process in the pharmaceutical industry; it ensures the uniformity of composition among all components by blending them; it is a vital step to achieve dose uniformity. To achieve optimal mixing, factors such as particle size, density, and shape need to be considered, as these can affect the mixing process. The main forces driving the mixing process are convection, diffusion, and shear. The goal of mixing is to reach a state known as a perfect mixture, where the concentration gradient of the drug is zero across the entire mixture. However, this state is unattainable in practice and instead, an acceptable mixture where the concentration gradient is minimal is desired. The rate of mixing follows first-order kinetics. The common equipment used for mixing is a rotating shell blender with and without an agitator, and a stationary shell. Difficulty in achieving blend uniformity of low-dose active ingredients with other excipients is often observed, which could be resolved by using customized approaches. Scale-up is also a significant consideration, and

K. Bansal
Formulation & Development, Amneal Pharmaceuticals Pvt. Ltd., Ahmedabad, India

A. B. Jindal (✉)
Department of Pharmacy, Birla Institute of Technology and Science, Pilani, Jhunjhunu, Rajasthan, India
e-mail: anil.jindal@pilani.bits-pilani.ac.in

A. B. Jindal (ed.), *Pharmaceutical Process Engineering and Scale-up Principles*, AAPS Introductions in the Pharmaceutical Sciences 13,
https://doi.org/10.1007/978-3-031-31380-6_2

trial-and-error or prior experience are the primary methods for achieving the desired outcome. Another approach, the Froude number method, is a mathematical model that has its own advantages and limitations in the scale-up of mixtures.

Keywords Mixing of powders · Blender · Driving forces in mixing · Blend uniformity

2.1 Introduction

Blending and Mixing can be defined as the reorientation of particles relative to one another in order to achieve uniformity (SUPAC: Manufacturing Equipment Addendum, 2014). Mixing is a process of random distribution of the particles to obtain a uniform blend with respect to the drug (Weidenbaum, 1958). In solid dosage form, the main objective of the mixing process is to obtain dose uniformity in the finished product. Dose uniformity is a very important criterion for the therapeutic effect of the pharmaceutical product. Improper mixing of the drug with the excipients could lead to the presence of the drug in the subtherapeutic/toxic range, which may result in no effect or toxic effect of the product. We can understand uniform mixing by the following simple example. Let's consider the mixing of yellow-coloured turmeric powder with white-coloured lactose. When can the blend of turmeric powder and lactose be called uniform? The uniformity of the blend can be confirmed by taking the samples from different locations and comparing the intensity of the yellow colour of the sample. When the intensity of the yellow colour in all the samples is the same, the blend is referred to as uniform.

2.2 Factors Affecting the Powder Mixing Process

Several factors, including particle size and particle size distribution, density, and shape, can affect the process of powder mixing (Ralf Weinekötter, 2013). Particle size plays a significant role in defining the uniformity of the powder blend. For instance, the mixing of coarse particles is easy as compared to the mixing of fine material. Fine particles exhibit high surface area and hence high surface free energy with high cohesive/adhesive forces compared to coarse particles, which may require additional shear force for uniform mixing. Another critical parameter that exerts a significant effect on mixing is the density of the powder. In general, densely packed powders exhibit difficulty in flow due to the formation of large aggregates. Such types of materials also show problems in mixing. Particle shape was also found to be very important for efficient mixing (Lachman et al., 1987). Those particles which are spherical in shape show better flow properties and hence better mixing as compared to the particles of any other geometry. The basic reason for the good flow properties of the spherical particles is minimum surface area and hence minimum surface free energy as compared to non-spherical shapes. In general, it can be concluded that any parameter which could affect the flow properties of the powder would exert a profound impact on efficient powder mixing.

2.3 Mechanism of the Mixing Process

The following three mechanisms accomplish the mixing process of the solids.

Convection Particles are reoriented in relation to one another as a result of mechanical movement, also known as paddle or plough mixing. In other words, it is the random movement of the group of adjacent particles from one place to the other in the mixture during the mixing process.

Diffusion Particles are reoriented in relation to one another when they are placed in random motion and interparticular friction is reduced as the result of bed expansion (usually within a rotating container); also known as tumble blending (SUPAC: Manufacturing Equipment Addendum, 2014). In other words, it is the random movement of the individual particle in the mixture from one location to the other during the mixing process.

Shear It is the separation of the cohesive particles due to the forces acting in the tangential direction on the surface of particles. It is a predominant mechanism of mixing fine and sticky materials. Shear force is provided by an agitating bar in the blender (Bauman et al., 2008).

2.4 Kinetics of the Mixing Process

The kinetics of the mixing process represents the rate at which the concentration of the drug changes at different locations in the blend. At the beginning of the mixing process, the change in the concentration of the drug is very fast, which tends to zero as the process continues. After a certain period, the drug concentration in the samples withdrawn from different locations in the blenders becomes almost equal, and the concentration gradient is negligible (Carter, 1996). It is important to note that the concentration gradient never becomes zero. The rate of change of concentration gradient in the blend follows first-order kinetics, which is the following equation can express

$$dC/dt = -Ae^{-kt}. \tag{2.1}$$

dC/dt = rate of change of the concentration gradient of the drug at the different locations in the blender

A = initial resistance in the mixing process
k = first-order rate constant, which depends upon the physical properties of the material and mixing mechanism of the blender

The state when the concentration gradient becomes zero is a perfect mixture. The mixing process continues for an infinite time to obtain the perfect mixture. Therefore, practically, no state of perfect mixture exists.

2.5 The Perfect and Acceptable Mixture

The perfect mixture is the state of the mixing process in which each sample is withdrawn from different locations in the blenders exhibiting the same amount of the drug (Herbert A. Liberman, 1990). The state of the perfect mixture can never be achieved during the mixing process. The following example can explain it. Let's assume that an equal number of white and black balls are present in a bag. If a person removes two balls from the bag several times, the possibility of one white and one black ball each time in the two balls removed from the bags is negligible. Similarly, obtaining a sample of the same composition from the different locations of the blender is negligible. As the state of the perfect mixture can never be achieved, consideration must be given to an alternative state in which the difference in the concentration of the drug present in each sample removed from the different locations of the blender is statistically insignificant. This state is known as an acceptable mixture.

As per current pharmaceutical practices, an acceptable mixture should conform to blend uniformity when samples from 10 different locations in the blend are tested. Mean of all results should be NLT 90.0% and NMT 110.0% (for the US market) and all individual results should be within ±10% of the mean. Relative standard deviation should be NMT 5.0%. Mean value of all results is stringent in some other countries like Canada, where NLT 95.0% to NMT 105.0% results are expected.

2.6 Mixing Equipment

Generally, the type of mixing equipment used for the mixing process can be classified into three broad categories: a) rotating shell blender with no agitator, b) rotating shell blender with agitator, and c) stationary shell with rotating mixing blades. The selection of the equipment depends upon the physical properties of the material which is to be processed (C.V.S Subramhanyam, 2001). Rotating shell blender with no agitator is suitable for the coarse powder while mixing fine and sticky materials can be accomplished by rotating the shell blender with agitators.

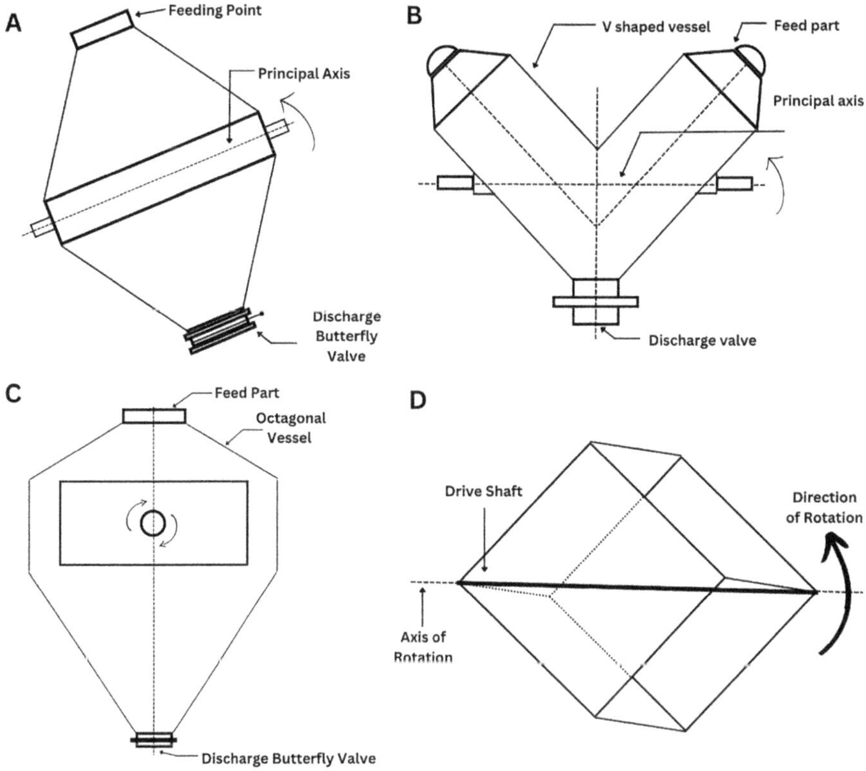

Fig. 2.1 Equipment used for mixing (**a**) double cone blender; (**b**) v-shaped blender; (**c**) octagonal blender; (**d**) cube shape blender

Equipment belonging to the first category is available in diverse geometric shapes, which include double cone, V- shape, octagonal, and cube. A schematic of a blender of different shapes is shown in Fig. 2.1.

Rotating shell blender with no agitator induces particle movement by rotating the entire body of the blender, which results in the tumbling of the powder within the blender. In such a blender, the mixing process is accomplished by the gravitational force which acts on the particles during tumbling in the blender. As it uses minimum energy for the mixing process, it is suitable for the mixing of friable materials. However, a rotating shell blender with no agitators cannot mix fines and sticky powder. It can be explained based on forces that act during the mixing process using such blenders. Fine and sticky materials exhibit high cohesive forces, which tend to form a loose aggregate of particles. To enable micro-mixing and obtain a uniform blend, it is essential to break these aggregates. To overcome the cohesive forces and present the fine and sticky materials, an additional shear force is required, which is absent in such blenders. In a rotating shell blender with an agitator, the mixing process depends upon the gravitational force, which cannot overcome the cohesive forces in fine and sticky materials. Therefore, rotating shell blenders with no agitator are unsuitable for mixing such materials.

Rotating shell blenders with no agitator offers several advantages, including being easy to install, clean and maintain. Moreover, it is available in wide sizes ranging from 3 L to 6000 L with geometric similarity in shape, which enables the scale-up process easy.

It also has a few limitations. For instance, segregation is a major concern after mixing powders using these blenders if the material has wide particle size distribution. Another limitation of these blenders is that serial dilution (i.e. geometric mixing) is required to mix low-dose active ingredients. Furthermore, it is also not suitable for the mixing of fines and sticky materials (Liberman, 1990).

In general, blenders are filled with 30–70% of the total volume of the blender. The fill level plays an important role in the mixing efficacy of the blenders. It has been observed that an increase in the fill level from 50 to 80% and an increase of 10 to 40 min in the blend time was seen in the production-size blender. In rotating shell blenders with no agitators, those particles present on the surface of the blend in the blender are in the cascade movement during the tumbling motion in the blender while the remaining blend is in a static state. As the fill level of the blender increases, the number of particles that are present on the surface of the blend increases (Liberman, 1990). An increase in the number of moving particles results in a decrease in the mixing time. Therefore, an increase in blend time was observed when the fill level was increased.

Another important factor that could exert a significant effect on the mixing efficacy is the speed of the blender. At a significantly low speed of the blender, the powder blend remains static in the blender. As the speed of the blender increases, powder starts moving alongside the walls of the blender; however, no mixing happens. When the speed of the blender is increased further, randomization of the particles is initiated under the influence of gravitational forces (Lachman et al., 1987). The speed at which the maximum random distribution of the particles is observed is known as the optimum speed of the blender. If the speed of the blender is increased beyond the optimum speed, it reaches the critical speed, and the powder blend starts rotating with the walls of the blender due to the centrifugal force generated at a higher speed. Therefore, the critical speed of the blender can be defined as the speed at which the powder blender just starts centrifuging. In general, the optimum speed is 80% of the critical speed of the blender. The critical speed of the blender depends upon the size of the blender, and the following equation can calculate it.

$$\omega = \left(\frac{g}{r}\right)^{1/2} \tag{2.2}$$

Where ω and r, are critical angular velocity and the radius of the blender, respectively, g is the gravitational acceleration.

As the size of the blender increases, critical speed decreases due to an increase in the centrifugal force, which results in a decrease in the optimum speed of the blender.

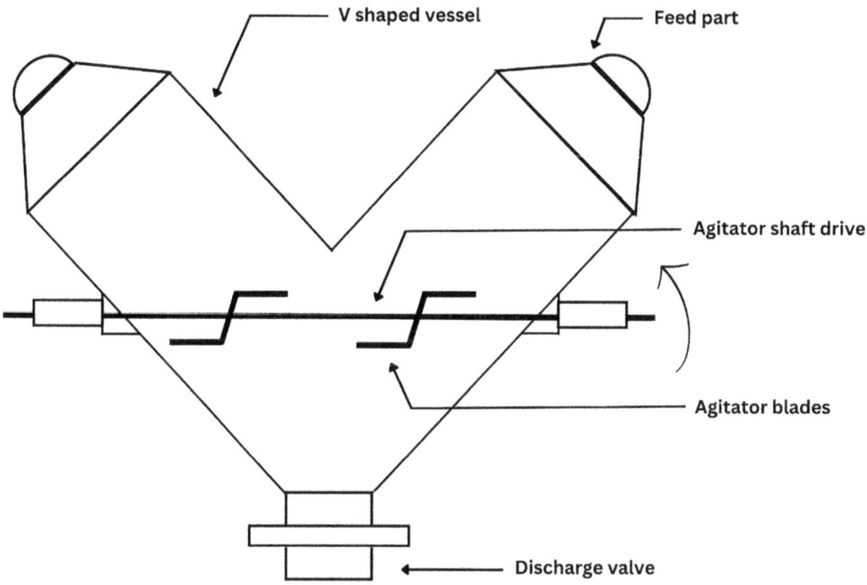

Fig. 2.2 V-shaped blender with agitator

Therefore, during the scale-up of the mixing process in a larger-size blender, the optimum speed decreases to obtain results similar to the lab-scale process.

Another important feature of the tumbling blenders is symmetry in design. The symmetrical design of the blender significantly increases the mixing time required to obtain a uniform blend. This is probably due to the limited movement of the material across the symmetric plane in the blender) (Liberman, 1990). Therefore, blenders are designed asymmetrically (e.g. slant cone) to avoid the formation of dead zones or facilitate the mixing process.

The second general category of mixing equipment is a rotating shell blender with a high-speed agitating bar (Fig. 2.2). The advantage of the presence of the agitator is that it can provide additional shear force during the mixing process, which is essential for the mixing of fines to overcome the cohesive forces. It is, therefore, the rotating shell blender with an agitator that is suitable for the mixing of cohesive materials. The addition of agitators in the tumbler mixers may also result in disadvantages. For instance, it is difficult to scale-up the process when an agitator is present in the blender due to the complexity of the design. Moreover, the addition of an agitator may result in the generation of more fines due to the increase in the attrition effect (Bauman et al., 2008).

The third category of mixing equipment involves equipment that has a fixed body with moving blades for the mixing process. The powder is confined to the fixed body of the mixer, and moving blades provide the energy for the mixing process. This category of mixer can be used for both wet and dry mixing. These mixers are available in many designs, including ribbon mixers, sigma blade mixers, planetary mixers, and conical screw mixers (Carter, 1996).

2.7 Mixing Problems

In the following section, mixing problems that are generally observed during the development of solid dosage forms are discussed.

Uniform Dispersion of the Low-Dose Active Ingredient with Other Excipients Obtaining desired blend uniformity of low-dose active ingredients with other excipients is significantly challenging (Dr. Bhawna Bhatt, 2008). The following strategies can be used to obtain the uniform dispersion of active ingredients with other excipients.

2.7.1 Milling of Active Ingredients

Milling of active ingredients can increase the blend uniformity significantly. The following hypothetical example can understand it. Let's consider the mixing of 10 g of the active ingredient with 100 g of diluent to obtain the ten dosage units. Assume that 10 g of the active ingredient and 100 g of diluent contain 10 and 100 particles, respectively. After the mixing, the theoretical weight of the mixture is 110 g which can be divided into ten dosage units of 11 particles each. To understand the impact of milling on blend uniformity, we will consider the following three scenarios.

Scenario 1

11 particles of mixture =	1 particle of active ingredient +10 particles of diluent
(1 dosage unit)	(Assay =100%)

Scenario 2

11 particles of mixture =	0 particles of active ingredient +11 particles of diluent
(1 dosage unit)	(Assay = 0%)

Scenario 3

11 particles of mixture =	2 particles of active ingredient +9 particles of diluent
(1 dosage unit)	(Assay = 200%)

In scenario 1, the assay of the dosage unit is 100% which may not be true for all the dosage units. There is a huge possibility that some of the dosage units may have either scenario two or scenario 3. Both scenario two and scenario 3 showed that the assay of the mixture varied from 0 to 200% due to the increase or decrease of 1 particle of the active ingredient in the mixture. It indicates the wide variation in the uniformity of the dosage units.

We will repeat the same exercise after milling the active ingredient to understand the effect of milling on the blend uniformity. Let's assume that 10 g of active ingredient contains 100 particles after the milling. After the mixing, the theoretical weight of the mixture is 110 g which can be divided into 10 dosage units of 20 particles each. To understand the impact of milling on the blend uniformity, the above exercise was repeated as follows.

Scenario 1

20 particles of mixture =	10 particles of active ingredient +10 particles of diluent
(1 dosage unit)	(Assay =100%)

Scenario 2

20 particles of mixture =	9 particles of active ingredient +11 particles of diluent
(1 dosage unit)	(Assay = 90%)

Scenario 3

20 particles of mixture =	11 particles of active ingredient +9 particles of diluent
(1 dosage unit)	(Assay = 110%)

After the milling, although 1 particle of active ingredient was increased or decreased in the mixture, the assay was varied from 90% to 110%. The variation in the assay of the dosage unit is significantly less as compared to the mixture which was obtained before milling.

The above example shows the significance of the milling in uniform dispersion of low dose active with the other excipients.

2.7.2 *Mixing of Poor-Flowing Cohesive Materials*

The cohesiveness of the particle is a surface phenomenon that is inversely proportional to the particle size. A decrease in the particle size results in an increase in the cohesive forces, which may lead to problems in the flow of the powder. During the mixing of the cohesive powders, an additional shear force is required to overcome the cohesive forces and increase the diffusive mixing of the material. It is suggested that using a tumbler mixer with agitators can improve the mixing of cohesive materials.

2.7.3 *Geometric Mixing*

Another strategy that can be used to obtain the uniform dispersion of the active ingredient is the serial dilution of the drug with the excipients. The geometric mixing method helps to achieve good blend uniformity for potent molecules having low dose.

2.8 Scale-Up Approaches for the Mixing Process

Scale-up of the mixing process is important for successfully translating the pharmaceutical product from the lab to the production scale. No mathematical model is available to predict the process parameter of powder blending during the scale-up. In the pharmaceutical industry, the successful scale-up of the mixing process largely depends upon the trial-and-error approach and prior experience of the product development scientist. In the following section, we will attempt to frame certain guidelines for the successful scale-up of the mixing process.

2.8.1 *Froude Number Approach*

Froude number is defined as the ratio of inertial and gravitational forces of the rotating blender and can be calculated from the following equation.

$$F = \frac{\mu^2 R}{g} \tag{2.3}$$

where and R is the rotational speed and radius of the blender, respectively. g is the gravitational acceleration.

Froude number is calculated and kept constant across the scale to determine the rotation speed and mixing time (Britannica, 2013). The use of the Froude number for calculating the scale-up parameter of the blending process is based on two assumptions. (a) Blenders which is used for the mixing process at different scales (e.g. laboratory scale, pilot scale, and production scale) should be geometrically similar, and (b) the number of revolutions of the blender during the mixing process should be kept constant (Levin, 2001).

Two blenders, laboratory scale and production scale, can be called geometrically similar if the ratio of corresponding angles and dimensions of the blenders are constant. However, collaborative efforts of manufacturing engineers and formulation scientists are required for the design of geometrically similar blenders for the successful scale-up.

In the Froude number approach, rotational speed has not been given importance. The number of revolutions is considered necessary. It should be kept constant for calculating mixing time as the degree of mixing increases with each revolution of the blender during the mixing process. The use of the Froude number for the calculation of scale-up parameters in the blending process has been illustrated in the following general problem.

Example 1 During the formulation of a solid oral dosage form, a blending process was optimized at a laboratory scale using a 5 L blender. The 5 L blender was operated at 15 rpm for 15 min to obtain the desired blend uniformity. Calculate the operating parameters, if the blending process was scaled up from 5 L to 25 L blender. Both the blenders are geometrically similar. [Assumption → Volume of blender is directly proportional to R^3]

According to the Froude number approach

$$F_5 = F_{25} \ldots 1.3$$

F_5 and F_{25} Froude number for 5lt and 25lt blender, respectively

$$(\mu_5)^2 R_5/g = (\mu_{25})^2 R_{25}/g$$

$$\mu_{25} = \mu_5 \ (R_5/R_{25})^{1/2}$$

$$R_5/R_{25} = \ (V_5/V_{25})^{1/3.} \ \left(\text{As } V \text{ is directly proportional to } R^3\right)$$

$$R_5/R_{25} = \ (5/25)^{1/3} = (1/1.70)$$

$$\mu_{25} = 15 \ (1/1.70)^{1/2}$$

$$\mu_{25} = 15 \times 1.30 = 19.5 \text{ rpm.}$$

In the case of a 5 L blender, the total number of revolutions is 225 and according to the scale-up principle total number of revolutions should be kept constant across the scale to calculate mixing time.

Therefore, mixing time for the 25 L blender = Number of revolution/rotational speed

$$= 225/19.5$$

$$= 11.5 \text{ min}$$

Practice Problems

(a) During the optimization of the mixing process using a tumbler mixer, if the volume of blender 1 is eight-fold higher than blender 2, what is the ratio of the rotational speed of blender 1 to blender 2? Both the blenders are geometrically similar. [Assumption → Volume of blender is directly proportional to R^3].

2.8.2 Limitations of the Froude Number Approach

Froude number is calculated based upon the assumption that the critical speed of the blender is equal to the optimum speed, which may not be true in practical conditions. Moreover, the Froude number approach can only be used when the blenders are geometrically similar across the scale, which may require maintaining a similar fill level (Levin, 2001). Maintenance of a similar fill level may result in an increase in the production time due to a decrease in the number of particles that are in the movement during the mixing process. Moreover, this approach can also not be used when blender geometry is changed during the scale-up (e.g. V-blender to double cone blender).

Although blending is an important unit operation in the development of solid oral dosage forms (e.g. tablets and capsules), there are no systematic guidelines available for the transfer of the blending process from lab-scale blender to the production-scale blender. Nevertheless, using geometrically similar equipment could solve the problem to a large extent, but procurement of dedicated equipment for each product is not possible for the industry. Furthermore, validation of the Froude number approach is also required in the industry to enhance its application in scale-up.

It has also been observed that large-size tumbler mixers with no high-speed agitators (e.g. 100 L or above) may not be able to produce the blend of desired uniformity. This could be because when the blender size increases, the energy required for the mixing per unit mass also increases. Two options are available to meet the high energy requirement in case of large batch size. Either increase the rotation speed of the blender or use a tumbler mixer with a high-speed agitator. The former approach (i.e. increase of the rotational speed of the blender) may not be advisable as the movement of the particles in large blenders may be ceased due to the generation of centrifugation forces. Therefore, in production, tumbler mixers with agitators are used to compensate for the increased energy requirement of the blender.

2.8.3 Tip Speed Approach

Tip speed approach is mainly used in the scale-up of rapid mixer granulator (dry and wet mixing) and blender (dry mixing). A tip speed is defined as the tangential velocity of an impeller at a point on its tip. The tip speed is a function of the RPM and diameter of the impeller.

The following formula is used to calculate the tip speed of an impeller.

TS = pi*D* RPM/60
Where TS is the tip speed (distance/second)
D is the diameter of the impeller (m or ft)
RPM is the rotations per minute

Example 2 During the formulation of a solid oral dosage form, a blending process was optimized at a pilot plant scale-up using a 1200 L V-blender. The 1200 L blender was operated at 12 rpm for 25 min to obtain the desired blend uniformity. Calculate the operating parameters, if the blending process was scaled up from 1200 L to 6000 L V-blender. Tip radius of 1200 L V Blender is 1000 mm and for 6000 L V-blender, it is 1600 mm.

Calculate the tip speed for 1200 L V-Blender at 12 rpm (*Tip speed Calculator*, 2009)

$$\text{Tip speed} = 0.628\ \text{m/s}$$

Now, calculate the tip speed for 6000 L V-blender, assuming rpm as 12:

$$\text{Tip speed} = 1.005\ \text{m/s}$$

Since the tip speed for 1200 L V-blender and 6000 L V-blender does not match, it is not recommended to operate the 6000 L V-blender at this rpm.

Now, change the rpm of 6000 L V-blender in such a way to match tip speed approx. 0.628 m/s:

At 8 rpm, the tip speed for 6000 L V-blender is calculated to be 0.670 m/s, while at 7 rpm it is 0.586 m/s and at 9 rpm, it is approx. 0.754 m/s.

Therefore, we can finalize tip speed of 0.670 m/s at 8 rpm, which is closer to tip speed of 1200 L V-blender. Minutes are kept same as that operated for 1200 L V-blender (i.e. 25 min).

2.8.4 Limitations of the Tip Speed Approach

Tip speed approach does not let the user to alter the time for which the machine is operated. Time of mixing could have significant impact on the degree of randomization achieved. Moreover, it is important to consider blend fill level and their particle size distribution for appropriate mixing of the blend, which is not taken into consideration by tip speed approach.

2.8.5 Number of Revolution Approach

This approach takes into consideration of time and rpm and calculates total number of revolutions. In case of scale-up, there may be cases where the same rpm is not possible to execute. In such cases, time is increased so as to match the total number of revolutions.

$$\text{Total number of revolutions} = \text{Time} \times \text{rpm}$$

Example 3 During the formulation of a solid oral dosage form, a blending process was optimized at a pilot plant scale-up using a 200 L Bin blender. The 200 L blender was operated at 15 rpm for 20 min to obtain the desired blend uniformity. Calculate the operating parameters, if the blending process was scaled up from 200 L to 1500 L Bin blender. Assume we have to operate 1500 L at 12 rpm as the machine is qualified in 8–12 rpm range.

Total number of revolutions in 200 L Bin Blender = 20 × 15 = 300.

Since we have to operate 1500 L Bin Blender at 12 rpm, we shall calculate time required to match number of revolutions = 300/12 = 25 min.

Therefore, 1500 L Bin Blender should be operated at 12 rpm for 25 min so as to match the number of revolutions of 200 L Bin Blender.

2.8.6 *Limitations of the Number of Revolution Approach*

Although matching of the number of revolutions is most simple and easy way to scale up and scale down and is used very widely in the pharmaceuticals system, this method has certain limitations. Number of revolutions approach does not take into consideration of the tip speed and it may eventually lead to a long time, in some cases, which may be 35 or 40 min of mixing. Such a long time of mixing may result in demixing.

2.9 Conclusion

Mixing is one of the preliminary steps in pharmaceutical formulation, and its proper execution remains a step of paramount importance. A proper understanding of the process is required to determine the factors responsible for its proper completion and the factors affecting it. Additional efforts have been undertaken to derive the mechanisms and mathematical models to achieve acceptable mixtures with negligible concentration variations. Finally, a proper understanding and optimization of the process is required for the proper scale-up.

References

Bauman, I., Ćurić, D., & Boban, M. (2008). Mixing of solids in different mixing devices. *Sadhana – Academy Proceedings in Engineering Sciences, 33*(6), 721–731. https://doi.org/10.1007/S12046-008-0030-5/METRICS

Britannica. (2013). *Froude number*. https://www.britannica.com/science/Froude-number

Carter, S. J. (1996). *Cooper and Gunn tutorial-pharmacy*. CBS Publishers.

Bhatt, B., & Agrawal, S. S. (2008). Mixing, Pharmaceutical Engineering, Edition 1, 2–24.

Lachman, L., Liberman, H. A., & J. L. K. (1987). *The theory and practice of industrial pharmacy*. Varghese Publishing House.

Levin, M. (Ed.). (2001). Pharmaceutical process scale-up. CRC Press.

Lieberman, H. A., Lachman, L., & Schwartz, J. B. (1990). *Pharmaceutical dosage forms: tablets, informa healthcare*. Marcel Dekker, INC.

Weinekötter, R., & Gericke, H. (2013). *Mixing of solids (Vol. 12)*. Springer Science & Business Media.

Subramhanyam, C. V. S. (2001). *Pharmaceutical engineering – Principles and practices*. Vallabh Publisher.

Tip Speed Calculator. (2009). https://www.simsite.com/tip-speed-calculator. (Accessed on 24 Dec 2022).

Weidenbaum, S. S. (1958). Mixing of solids. *Advances in Chemical Engineering, 2*(C), 209–324. https://doi.org/10.1016/S0065-2377(08)60229-X

Chapter 3
Rapid Mixer Granulator

Rikin Patel, Siddharth Salve, Dhwani Rana, Amit Sharma, K. Bharathi, Sagar Salave, and Derajram Benival

Contents

Abstract For manufacturing solid oral dosage forms, such as tablets and capsules, good flow of granules is crucial. Granulation of poorly flowable powder can improve not only the flow but also helps in maintaining the content uniformity required to produce a quality drug product. Among various granulation techniques, wet granulation using rapid mixer granulator is one of the most widely used unit operations in the manufacturing of solid oral dosage forms. This chapter will cover the types of granulation processes and discuss in detail the wet granulation process using rapid mixer granulator. The goal is to provide an overview of the wet granulation process using rapid mixer granulator.

R. Patel
Intas Pharmaceuticals Ltd., Ahmedabad, Gujarat, India

S. Salve · D. Rana · A. Sharma · K. Bharathi · S. Salave · D. Benival (✉)
National Institute of Pharmaceutical Education and Research-Ahmedabad (NIPER-A) An Institute of National Importance, Government of India, Department of Pharmaceuticals, Ministry of Chemicals and Fertilizers, Gandhinagar, Gujarat, India
e-mail: derajram@niperahm.res.in

A. B. Jindal (ed.), *Pharmaceutical Process Engineering and Scale-up Principles*, AAPS Introductions in the Pharmaceutical Sciences 13,
https://doi.org/10.1007/978-3-031-31380-6_3

Keywords Wet granulation · Dry granulation · Rapid mixer granulator · Impeller · Chopper

Abbreviations

API	Active pharmaceutical ingredient
ASTM	American Society for Testing and Materials
CPP	Critical process parameter
CQA	Critical quality attribute
DEM	Discrete element modelling
PBM	Population balance modelling
QbD	Quality by design
RMG	Rapid mixer granulator
RPM	Revolutions per minute

3.1 Introduction

Tablets and capsules are the two major solid oral dosage forms which offer several advantages, including easy administration of unit dose by the patient, optimizing the absorption rate of the drug, masking the unpleasant taste of a therapeutic agent, and controlling the rate as well as the site of drug absorption. Different kinds of tablets like simple uncoated tablets, coated tablets, buccal tablets, chewable tablets, sublingual tablets, multilayer tablets, orodispersible tablets, osmotic tablets, and extended-release tablets are developed. Similarly, capsules of various sizes and various release profiles are also developed. Mainly, it is the dose and properties of the active pharmaceutical ingredient (API) which influence the design of a tablet and capsule formulations. Most of the APIs when used alone, demonstrate either poor compression properties or poor disintegration or poor dissolution or a combination of these properties. To overcome these limitations of the API, different kinds of other materials, called as excipients, are included in tablet and capsule formulations. Since these excipients control the biological performance of the dosage form and therefore, such excipients are called functional excipients. In addition to these functional excipients, a tablet/capsule formulation can also contain non-functional excipients which are used to facilitate the manufacturing of the dosage forms.

3.2 Need of Powder Granulation

Probably the simplest approach to formulating a tablet is to compress a powder blend obtained by uniform mixing of API and excipients. However, a granulation step is often necessary for three main reasons: to get good flow during compression in a high-speed rotary tablet press, to improve compactability during compression, and to maintain content uniformity from tablet to tablet in a batch.

3.3 Types of Granulations

Granulation is a unit operation which involves conversion of powder blend into granules, or physical aggregates of powders. It is important to note that in the granules, original particles of different powders that participated during the granulation process can still be identified. Since multiple excipients of different functionalities are used in a tablet formulation, granulation of such powder blend commonly involves adhesion of multiple particles of more than one type of excipient powder. Granulation can be done with or without the use of water or any other organic solvent, such as isopropyl alcohol. Based on this, granulation is usually classified into two categories.

3.3.1 Dry Granulation

Dry granulation, which is also known as dry compaction, does not involve the use of liquid (Kleinebudde, 2004) and it is used mostly for heat or moisture-sensitive products. In dry compression, API and excipients are first mixed in a blender and the powder blend so obtained can be further processed either by slugging method or by roller compaction. Slugging method involves compression of powder blend in the die cavity to form large slugs (e.g. 20 mm diameter) in a heavy-duty tablet press. The slugs so obtained are milled using a suitable mill, such as clit mill, to obtain granules of desired size. In roller compaction method, powder blend is passed through the gap present between the two counter-rotating rollers and sheets of compressed material which are called as ribbons or compacts is obtained. These ribbons are then subjected to milling, similar to that of slugs to obtain granules of desired size. In the authors experience, slug preparation using heavy-duty tablet press is more widely used in the industry in comparison to roller compaction method. This popularity is based on the cost and space considerations. Heavy-duty tablet press is readily available in all the pharmaceutical industries irrespective of the method of granulation (i.e. wet granulation or dry granulation for compression of granules into tablets). Hence, no separate investment is needed whenever an API demands dry granulation. However, in the case of roller compaction method, new investment in terms of cost as well as space would be required. In addition, the number of products manufactured by dry granulation in comparison to wet granulation and direct compression are significantly less, and hence, considerations are also needed in terms of periodic maintenance of roller compactor.

3.3.2 Wet Granulation

As mentioned earlier, wet granulation involves the use of water or any other organic solvent and there are mainly three ways by which wet granulation can be performed. The first variant of wet granulation, which is also the classical traditional wet granulation unit operation, involves first mixing of API and excipients using a V-shaped mixer. The resulting premixed blend is then transferred into a traditional low-shear granulator and a binder solution is then added slowly into this powder blend under a mechanical shear until a desired granule size is obtained. These wet granules are then sieved through appropriate mesh size and dried in a tray dryer. Dried granules are subjected to milling to obtain granules of desired size which are then mixed with lubricant to obtain the lubricated blend which is now ready for compression into tablets.

The second variant of wet granulation involves mixing of API and excipients directly into a high shear granulator and then the binder liquid is added to obtain the wet granules. Since these granules are obtained in a very short time and therefore high shear granulator is popularly known in the industry as Rapid Mixer Granulator (RMG). Obtained wet granules may or may not be subjected to sizing depending on the excipient properties, API dose, and physicochemical properties. Uniform wet granules are then dried in fluid bed dryer and dried granules are processed in the same way as mentioned above.

The third variant of wet granulation involves first mixing of API and other excipients in a V-shaped or any other blender and loading this premixed blend into a fluidized bed granulator. A binder liquid is then sprayed to obtain granules of the desired size. In a fluidized bed granulator, drying also occurs simultaneously.

3.3.3 Hot Melt Granulation

Hot melt extrusion technology has numerous applications in drug delivery; however, to the best of the author's knowledge, currently available marketed drug products use its bioavailability-enhancing potential. In a hot melt extruder, extrudes are first milled into particles of desired size and the resulting powder is then processed in the same way as that of dried milled granules from wet granulation technique or milled granules from dry granulation technique. Although spray drying technique can be used to produce free-flowing powder, the technique is primarily reserved for specialized applications in drug delivery rather than using it solely to produce a free-flowing powder. Various granulation options are shown in Fig. 3.1.

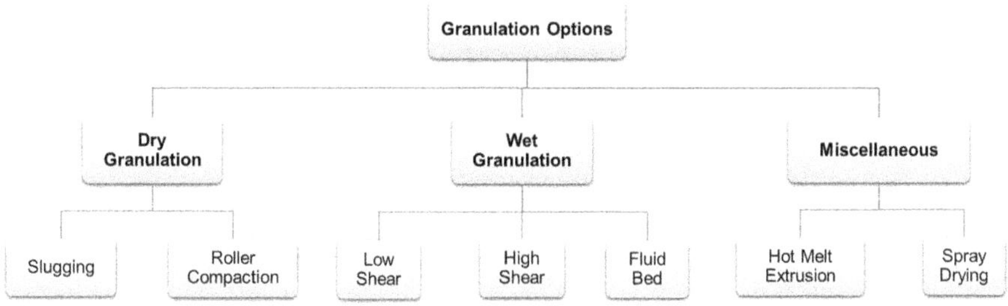

Fig. 3.1 Various types of granulation

The selection of a specific granulation technique depends on many factors including the nature of API and dose, desired characteristics of the granules such as particle size distribution and density of the granules, dissolution performance of the tablets, and cost consideration (availability of equipment at the manufacturing site).

Among all the granulation methods, wet granulation is the most widely used. The process involves agglomeration of different powders of various excipients and API by the addition of a granulating liquid under mixing. As the granulating fluid is added, binding of different particles together starts under the influence of capillary and viscous forces. As the drying of granules proceeds, these forces gradually start weakening with the formation of permanent bonds and in dried granules, permanent forces completely replace these forces (Iveson et al., 2001). The detailed mechanism involved in wet granulation is discussed in Fig. 3.1.

3.4 Mechanism of Granule Formation

Granule formation in a wet granulation process usually involves three stages (Ennis & Litster, 1997):

- Wetting and nucleation
- Consolidation and growth
- Attrition and breakage

3.4.1 Wetting and Nucleation

In this stage, which starts with the addition of binder liquid, different particles of various excipients and API, which are initially present as dry powder blend, come in contact with each other and result in the generation of nuclei granules. In the granulation process, this is an important stage. It is to be noted that this stage is rarely identified and difficult to separate from the effects of other stages of granulation such as coalescence and attrition. Material attributes as well as process variables affect this stage of granulation. Material attributes include appropriate selection of binder, which can provide efficient wetting depending on the contact angle it makes with the powder blend. Efficient mixing of powder and binder liquid is a process variable which is affected by binder liquid addition rate as well as spray characteristics. Uncontrolled binder liquid addition rate or poor spray characteristics may result in granules significantly different in properties from lot to lot.

3.4.2 Consolidation and Growth

This stage of granulation involves intergranular collisions, collision between granules and powder which is yet to be wetted or partly wetted as well as collision of granules with the wall of equipment and lead to compaction and growth of granules. When two large granules collide with each other and stick together, it results in granule growth, and, traditionally, this process is known as 'coalescence'. On the other hand, when fines get deposited on the surface of granule, then this phenomenon is called 'layering'. It is to be noted that various processes mentioned above responsible for the growth of granules can take place simultaneously during wetting as well as nucleation stage and may continue even after completion of binder liquid addition. Further, factors such as mechanical strength of granules as well as the presence of binder liquid at the surface of granules participating in collision activity which will ultimately decide whether a collision will result in permanent coalescence of two granules or not. Existence of different states of liquid saturation in granules is reported and this includes (Newitt & Conway-Jones, 1958):

- Pendular state: In this state, particles are held together by liquid bridges at their contact points (pendular bonds).
- Capillary state: This state prevails when saturation of granule with binder liquid occurs. In this state, all the voids get filled with binder solution and the binder solution present on the granule surface is drawn back into the pores due to the capillary action.
- Funicular state: The transition state between the pendular and capillary state in which the voids are not completely saturated with binder liquid is known as funicular state.
- Droplet state: This state arises when the particles are held within or at the surface of a binder liquid drop.
- Pseudodroplet state: As the name suggests, this state occurs when the unfilled voids remain trapped inside the droplet.

It is important to note that as the granulation process progresses, it is possible that binder liquid saturation state of the granules may shift from the pendular state to droplet state. This shift is either due to continuous addition of liquid binder solution or due to consolidation of granules which results in decrease in the granular porosity (Iveson et al., 2001).

3.4.3 Attrition and Breakage

As the name suggests, in this stage, wet or dried granules start breaking which may be due to shear, impact, or compaction in the granulator or even during the handling of the granules. Breakage of wet granules influences the final granule size distribution, particularly in RMG, while in the case of dry granules, attrition causes generation of excessive fines.

3.5 Rapid Mixer Granulator

When it comes to wet granulation process, then probably rapid mixer granulator (RMG) would the one which is one of the most widely used equipment. RMG is used to obtain dense granules having better flow and good compaction properties. It has been reported that RMG provides better control on content uniformity. Further, as the wetting of API is better in RMG, therefore, in some cases, it has been used to improve oral bioavailability of drug, including mitigation of food effect (Pandey et al., 2014; Pandey et al., 2012). During drug development, the manufacturing process of API undergoes significant changes to control the level of impurity. In addition, there is also an effect of scale-up process. During this journey from small-scale manufacturing to controlling the level of impurities to scale-up, changes in API powder properties are expected and RMG being a very robust granulation process, can overcome this issue related to variations in API powder properties.

The majority of RMGs comprise a cylindrical or conical mixing bowl, a three-blade impeller, a chopper, a motor to drive impeller and chopper, and a port for unloading the wet granules. Further, the bowl may be jacketed to control product temperature by circulating hot or cool liquids. The function of impeller is to mix the powders and to distribute the binder solution. Generally, it rotates from 100 to 500 revolutions per minute (rpm) (Murugesu, 2016). The chopper is usually rotating from 1000 and 3000 rpm and it helps in the breakdown of large agglomerates into granules. RMG having interchangeable bowls of different capacities such as 2 L, 5 L, and 10 L are also available for development use purposes. This type of RMG with interchangeable bowl capacity helps in understanding the effect of process scale-up. A schematic diagram of RMG and a picture of interchangeable RMG bowls is shown in Fig. 3.2. RMG is very popular in pharmaceutical industries in the manufacturing of solid oral formulations. As a result, a relatively large number of vendors offer RMG.

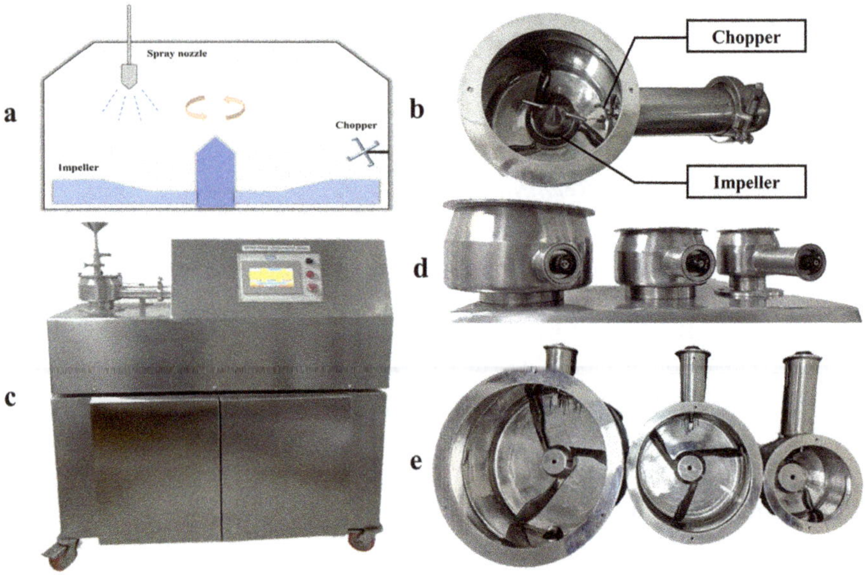

Fig. 3.2 (**a**) Schematic diagram of a vertical shaft bottom mounted RMG; (**b**) picture of RMG bowl showing impeller and chopper position; (**c**) picture of RMG; (**d**) interchangeable RMG bowls (2 L, 5 L, and 10 L: side view); and (**e**) interchangeable RMG bowls (2 L, 5 L and 10 L: top view)

A typical wet granulation process using RMG involves the following steps:

(a) Loading of API(s) and various excipients into the mixer bowl: Before loading API and excipients, it is necessary that all the raw materials must be sifted using appropriate sieve. This step helps in breaking down the agglomerates present in individual raw materials. Generally, excipients are sieved through the American Society for Testing and Materials (ASTM) # 40 mesh. However, in the case of API, selection of mesh size generally depends on the dose of API. If the dose of API is high, then it is fine to sift the API using the same mesh size as that of excipients (i.e. ASTM # 40). For low dose API, to improve the content uniformity, generally, micronized API is used (e.g. d90 < 5 microns, i.e. 90% of the API particles are less than 5 microns). This micronization of API is generally performed using fluid energy mill (jet mill). During this micronization, static charges are developed on the surface of particles and hence, such particles undergo agglomeration. If these agglomerates are not broken down before loading into RMG bowl, then this will result in content uniformity issue as RMG impeller or chopper would not be able to break such small agglomerates. Hence, for such low dose micronized API, it is generally advisable to use ASTM #60 or finer mesh to avoid content uniformity issue.
(b) Dry mixing of the loaded powders to obtain powder blend: In a typical RMG, uniform mixing can be achieved in 5–15 mins. Mixing uniformity can be confirmed by taking samples from the RMG bowl from different locations

including top, bottom, as well as near to the centre of the impeller. In addition, such sampling should be done at different time points. All these studies can be done during product development and should be verified during scale-up as impeller speed at R & D and at production scale may be different and this can affect the uniform mixing. The decision to use chopper (on, off, or intermittent on) during dry mixing should be based on the properties of the powder mixture.

(c) Addition of binder solution onto the moving powder bed: Binder solution can be poured directly/added through peristaltic pump or can be sprayed using a spray nozzle. Generally, binder solution is prepared by dissolving binder excipient in water or other solvent depending on the API/excipient requirement. Sometimes, water alone is used as binding liquid and binder excipients are included in the dry mix. However, dissolving binder in a water/organic solvent is a more widely used approach rather than adding it in the dry mix. Binder solution is added when impeller is running. However, chopper can be kept on or off depending on the product requirement. Use of chopper during binder solution addition can avoid generation of large agglomerates by breaking down the wet lumps.

(d) Once the binder solution addition is complete, rotation of impeller is still continued and now running of chopper is also required to perform kneading or wet massing.

(e) Discharge of wet granules from the RMG bowl followed by wet sieving (it is to be noted that sieving step is not mandatory). For small batch size, such as 500 G, wet granules are generally hand sieved through ASTM # 8 or 10; however, for large batch size, a mill is used. Milling of wet granules to obtain uniform-sized granules has some important implications. If larger wet granules are not milled before loading into fluid bed dryer, then there is a higher risk of "case hardening", in which such granules generally dry from surface while core still remains wet only. In fluidized bed dryer, larger granules fluidize differently in comparison to small size granules.

(f) Drying of wet granules can be performed in a fluid bed dryer. Traditionally, tray dryer was used for this purpose.

(g) Milling and sizing of dried granules. Milled granules are then further subjected to further processing such as lubrication, compression, or filling into capsules.

From the Quality by Design (QbD) perspective, various primary process variables and critical process parameters (CPPs) include:

- Batch size
- Impeller speed
- Binder solution addition method and rate
- Chopper speed and
- Wet massing time

3.5.1 End Point Determination in RMG

The end point in wet granulation using RMG is usually defined as a target mean granule size or granule size distribution. Monitoring mixer power consumption (Holm et al., 1985; Levin, 2006), impeller torque and torque rheometer (Corvari et al., 1992; Levin, 2006; Rowe & Sadeghnejad, 1987), acoustic emission (Whitaker et al., 2000), near-infrared moisture sensor (Miwa et al., 2000; Rantanen et al., 2005) and focused beam reflectance (Huang et al., 2010) are some of the approaches explored to determine the end point of wet granulation in RMG (Dun & Sun, 2019).

Advantages of wet granulation using RMG

- Short processing time
- Uniform granule size distribution
- Greater densification and reduced granule friability
- Different kinds of formulations such as immediate-release and sustained-release products can be processed
- Can be used to process highly cohesive raw materials
- Generally require lesser binder solution in comparison to low-shear granulator

Major limitations of RMG

- Narrow range of operating conditions
- Not suitable for those API which may undergo polymorphic transitions due to high shear
- Reduced granule compressibility relative to low-shear granulation

3.6 Types of RMG

Based on the impeller shaft orientation, batch RMG can be classified into two main types: vertical shaft and horizontal shaft. In the case of vertical shaft RMG, an impeller is either on the bottom or top-mounted while chopper is located at the side of the bowl. It is important to note that in this type of RMG, impeller facilitates powder bed to rotate under the spray zone of the binder solution. On the other hand, in the case of horizontal shaft RMG, impeller lifts and distributes the powder under the spray zone of binder solution and chopper is located near the bottom of the bowl. It is believed that in vertical shaft RMG, granules are thrown extensively against the wall due to centrifugal forces, and hence, granules from this type of RMG may be much more denser in comparison to horizontal shaft RMG (Kristensen & Schaefer, 2008).

3.7 Scale-Up of Wet Granulation Process in RMG

For cost-efficient development, generally, any product is developed at a smaller scale (few hundred grams for solid/semi-solid product to a few hundred mL for liquid product). Once an optimized formulation composition is obtained, then the formula is considered for scale-up and later for commercialization. However, it has been observed that critical quality attributes (CQAs) of a drug product may get adversely affected when the scale of manufacturing changes, particularly, when there are critical process parameters (CPPs) which change with the scale. For the smooth scale-up of wet granulation process using RMG, traditionally, dimensional analysis has been applied. According to this, in dynamically similar systems, dimensionless numbers required to describe the process have the same numerical value. Most commonly used dimensionless number for wet granulation process using RMG are Newton number, Froude number, and Reynolds number (Levin, 2006) (Table 3.1).

The following variables of RMG should be considered during the scale-up process.

3.7.1 Height of the Raw Materials in the Bowl

Bowl occupancy (in percentage) should be kept similar across various scales. Overall energy being used in the granulation process will be dependent on the fill height in the bowl (Verma et al., 2019).

3.7.2 Binder Solution Spray/Addition Rate

The binder solution should be added slowly to avoid localized over-wetting. A spray using pressure nozzle can be beneficial in ensuring consistency in binder solution addition rate. It is also suggested that binder solution addition time should be comparable at various scales.

Table 3.1 Dimensionless numbers for RMG

Dimensionless number	Explanation
Power number	It pertains to the hindrance to the movement of the impeller due to the particles and their inertia, and it denotes the energy consumed by the impeller during mixing and granulation.
Froude number	It is defined as the ratio of inertia to the gravitational forces and represents the dynamic similarity of the process across the different scales.
Reynolds number	It represents the relationship between the inertial forces and viscous forces in fluid flow, and it is used to classify the type of flow behaviour

3.7.3 Chopper Speed

Normally, chopper speed has no significant effect on granule size. Chopper helps in breaking down the large agglomerates into small granules which can be further available for growth through 'coalescence' or 'layering' mechanism during the consolidation and growth stage of granulation. If the formulation components have the propensity to undergo large agglomerate formation, then speed and timing of chopper use during binder addition can have a significant role to play.

3.7.4 Impeller Speed

It is the impeller in RMG which mainly decides the quality of wet granules. The particle force and velocities are majorly dependent on the speed of impeller in any RMG design (Tao et al., 2015). In the old generation RMG, it was not possible to set the desired impeller speed; however, new generation RMG are equipped with variable frequency drive, and as a result, it is possible to set the exact RPM required for a particular product depending on the fill height/batch size. Power law correlation is the most commonly used scaleup rule for impeller speed and it is governed by the following equation:

$$\frac{n_2}{n_1} = \left(\frac{d_1}{d_2}\right)^n \tag{3.1}$$

In this, n_1 and d_1 are impeller speed (in RPM) and diameter for RMG scale 1 while n_2 and d_2 are impeller speed (in RPM) and diameter for RMG scale 2. According to this equation, if the value of n is 1, then it suggests to maintain a constant impeller tip speed across scales. On the other hand, if the value of n is 0.5, then it suggests to maintain a constant Froude number across scales. In constant impeller tip speed across scales, it is assumed that the granulation is based upon the impaction of the impeller blade against the granules and it does not consider forces due to the fill weight of materials and centripetal/centrifugal forces. Constant Froude number across scales, on the other hand, takes into consideration the greater forces generated in smaller-scale RMG, which is due to the effect of the bowl wall. The range of Froude numbers is much higher in fast running small-scale RMG machines as the Froude number is related to the square of the impeller speed (in rpm). Therefore, before choosing the speed (RPM) of the small RMG, it is paramount to know the size of the granulator in which the scale-up batch would be manufactured.

3.8 Modelling and Simulation in RMG

Modelling and simulation in wet granulation using RMG can potentially be used for improving the understanding of the effect of process parameters on critical quality attributes (CQAs) of the drug product. It can also help during scale-up of the process. Different modelling approaches used for the RMG process include discrete element modelling (DEM), regime-map-based models, population balance modelling (PBM), PBM coupled with computational fluid dynamics, PBM with compartmental model and DEM, PBM coupled with DEM, and PBM coupled with volume of fluid models (Immanuel & Doyle, 2005; Nakamura et al., 2013; Ramachandran et al., 2009; Kumar et al., 2013; Gantt & Gatzke, 2005; Chaudhury et al., 2014; Badawy & Pandey, 2017).

3.9 Conclusion

This chapter has briefly discussed various granulation techniques. Wet granulation using rapid mixer granulator was discussed in detail. General steps followed in a typical wet granulation process using RMG were also discussed. Further, the types of RMG were also reviewed along with the scale-up considerations. In the end, modelling and simulation in RMG were also briefly discussed.

References

Badawy, S., & Pandey, P. (2017). Design, development, and scale-up of the high-shear wet granulation process. In Y. Qui, Y. Chen, Z. GGZ, L. Yu, & R. V. Mantri (Eds.), *Devloping solid oral dosage forms pharmaceutical theory and practice* (pp. 749–776). Elsevier.

Chaudhury, A., Barrasso, D., Pandey, P., Wu, H., & Ramachandran, R. (2014). Population balance model development, validation, and prediction of CQAs of a high-shear wet granulation process: Towards QbD in drug product pharmaceutical manufacturing. *Journal of Pharmaceutical Innovation, 9*, 53–64.

Corvari, V., Fry, W. C., Seibert, W. L., & Augsburger, L. (1992). Instrumentation of a high-shear mixer: evaluation and comparison of a new capacitive sensor, a watt meter, and a strain-gage torque sensor for wet granulation monitoring. *Pharmaceutical Research, 9*, 1525–1533.

Dun, J., & Sun, C. C. (2019). Structures and properties of granules prepared by high shear wet granulation. In A. S. Narang & B. SIF (Eds.), *Handbook of pharmaceutical wet granulation theory and practicein a quality by design paradigm* (pp. 119–147). Elsevier.

Ennis, B. J., & Litster, J. D. (1997). Particle size enlargement. In R. H. Perry & D. W. Green (Eds.), *Perry's chemical engineers' handbook* (pp. 20–89). McGraw-Hill.

Gantt, J. A., & Gatzke, E. P. (2005). High-shear granulation modeling using a discrete element simulation approach. *Powder Technology, 156*, 195–212.

Holm, P., Schaefer, T., & Kristensen, H. G. (1985). Granulation in high-speed mixers part VI. Effects of process conditions on power consumption and granule growth. *Powder Technology, 43*(3), 225–233.

Huang, J., Kaul, G., Utz, J., et al. (2010). A PAT approach to improve process understanding of high shear wet granulation through in-line particle measurement using FBRM C35. *Journal of Pharmaceutical Sciences, 99*(7), 3205–3212.

Immanuel, C. D., & Doyle, F. J. (2005). Solution technique for a multi-dimensional population balance model describing granulation processes. *Powder Technology, 156*, 213–225.

Iveson, S. M., Litster, J. D., Hapgood, K., & Ennis, B. J. (2001). Nucleation, growth and breakage phenomena in agitated wet granulation processes: A review. *Powder Technology, 117*(1–2), 3–39.

Kleinebudde, P. (2004). Roll compaction/dry granulation: Pharmaceutical applications. *European Journal of Pharmaceutics and Biopharmaceutics, 58*(2), 317–326.

Kristensen, H. G., & Schaefer, T. (2008). Granulation: A review on pharmaceutical wet-granulation. *Drug Development and Industrial Pharmacy, 13*(4–5), 803–872.

Kumar, A., Gernaey, K. V., Beer, T., & De, Nopens, I. (2013). Model-based analysis of high shear wet granulation from batch to continuous processes in pharmaceutical production – A critical review. *European Journal of Pharmaceutics and Biopharmaceutics, 85*, 814–832.

Milling, M. B. (2016). In L. L. Augsburger & S. W. Hoag (Eds.), *Pharmaceutical dosage forms – Tablets unit operations and mechanical properties* (pp. 175–193). Taylor and Francis.

Miwa, A., Yajima, T., & Itai, S. (2000). Prediction of suitable amount of water addition for wet granulation. *International Journal of Pharmaceutics, 195*(1–2), 81–92.

Nakamura, H., Fujii, H., & Watano, S. (2013). Scale-up of high shear mixer-granulator based on discrete element analysis. *Powder Technology, 236*, 149–156.

Newitt, D., & Conway-Jones, J. (1958). A contribution to the theory and practice of granulation. *Transactions. Institute of Chemical Engineers, 36*(6), 422–442.

Pandey, P., Sinko, P. D., Bindra, D. S., Hamey, R., Gour, S., & Vema-Verapu, C. (2012). Processing challenges with solid dosage formulations containing vitamin E TPGS. *Pharmaceutical Development and Technology, 18*, 296–304.

Pandey, P., Hamey, R., Bindra, D. S., et al. (2014). From bench to humans: Formulation development of a poorly water soluble drug to mitigate food effect. *AAPS PharmSciTech, 15*, 407–416.

Ramachandran, R., Immanuel, C. D., Stepanek, F., Lister, J. D., & Doyle, F. J. (2009). A mechanistic model for breakage in population balances of granulation: Theoretical kernel development and experimental validation. *Chemical Engineering Research and Design, 87*, 598–614.

Rantanen, J., Wikström, H., Turner, R., & Taylor, L. S. (2005). Use of in-line near-infrared spectroscopy in combination with chemometrics for improved understanding of pharmaceutical processes. *Analytical Chemistry, 77*(2), 556–563.

Rowe, R. C., & Sadeghnejad, G. R. (1987). The rheology of microcrystalline cellulose powder/water mixes — Measurement using a mixer torque rheometer. *International Journal of Pharmaceutics, 38*(1–2), 227–229.

Tao, J., Pandey, P., Bindra, D. S., Gao, J. Z., & Narang, A. S. (2015). Evaluating scale-up rules of a high-shear wet granulation process. *Journal of Pharmaceutical Sciences, 104*(7), 2323–2333.

Verma, R., Patil, M., & Paz, C. O. (2019). Current practices in wet granulation-based generic product development. In A. S. Narang & B. SIF (Eds.), *Handbook of pharmaceutical wet granulation theory and practicein a quality by design paradigm* (pp. 203–259). Elsevier.

Wet, L. M. (2006). Granulation: End-point determination and scale-up. In J. Swarbrick (Ed.), *Encyclopedia of pharmaceutical technology* (pp. 4078–4098). Taylor and Francis.

Whitaker, M., Baker, G. R., Westrup, J., et al. (2000). Application of acoustic emission to the monitoring and end point determination of a high shear granulation process. *International Journal of Pharmaceutics, 205*, 79–91.

Chapter 4
Fluid Bed Processing Technology

Vishvesh M. Joshi and Anil B. Jindal

Contents

Abstract Fluid bed processing technology is widely used in the pharmaceutical industry for the processing of multiparticulates. The chapter describes the theory of fluidization followed by the components of a fluidized bed processor and their functioning. Further, the formulation, process and equipment-related factors that affect product quality are described in detail. Being a complex process, the process scale-up principles of the processor are discussed.

Keywords Fluid bed processor · Wurster process · Fluidization of the material · Fluid bed coating

V. M. Joshi
Alembic Labs LLC, West Caldwell, NJ, USA

A. B. Jindal (✉)
Department of Pharmacy, Birla Institute of Technology and Science, Pilani, Jhunjhunu, Rajasthan, India
e-mail: anil.jindal@pilani.bits-pilani.ac.in

A. B. Jindal (ed.), *Pharmaceutical Process Engineering and Scale-up Principles*, AAPS Introductions in the Pharmaceutical Sciences 13,
https://doi.org/10.1007/978-3-031-31380-6_4

4.1 Introduction

Fluid bed processing is extensively utilized in the industry for the manufacturing of solid oral dosage forms. It can be used to accomplish different unit operations, including granulation, drying, and palletization. It can perform three-unit operations, namely mixing, granulation and drying, in a single unit, which decreases the processing time significantly and improves the economy of the process (Alderborn & Aulton, 2002). Moreover, in the fluidized bed processor, particles are in the fluidized state during the drying, which increases the drying rate enormously due to the exposure of the large surface area of particles to heat and binder solution. The granules obtained by the fluid bed granulation process are more compressible than the high shear granulation process. It is a complex process due to the interdependence of the process parameters which makes the scale-up of the process more difficult (Wiling & Levin, 2001). Challenges present in the fluid bed granulation process are highlighted below

(a) Nonuniform particle size distribution of the granules
(b) Lump formation during granulation
(c) Clogging of the spray nozzle due to spray drying of the binder solution
(d) Poor blend uniformity due to the entrainment of the fines.

4.2 Fluidization Theory

In the fluidized bed process, particles remain fluidized in the presence of an airstream flowing in the upward direction. Particles start fluidizing when the air velocity is equal to the minimum fluidization velocity (i.e. air velocity is greater than the settling velocity and less than the entrainment velocity). When the air velocity is less than the settling velocity, the particle bed is in a static state. As the air velocity is increased, an upward movement of the particles is initiated, and particles start suspending in the air. During the air suspension, the upward-moving airstream exerts a drag force on the particles, which is balanced by the gravitational force acting in the downward direction. A particle remains fluidized when the drag force of the air on the particle is equal to the apparent weight of the particles (Carter, 1996; Alderborn & Aulton, 2002). As the fluidized bed height is increased in the equipment, pressure drop increases. The fluidization pattern of the particles in the equipment depends upon the different forces acting on the particle during the process (Stenlake, 1968). Van der Waals forces play a predominant role in the process although electrostatic forces also cannot be ignored. For instance, in the case of fines, the presence of cohesive forces may result in the fluidization of the particles as aggregates instead of individual particles. In a fluidized powder bed, pressure drop increases with an increase in the air velocity until the velocity has reached the minimum fluidization velocity. It is the state pressure drop which is equal to the force experienced per unit area of the particle. Pressure drop remains constant until the air velocity is less than the entrainment velocity of the particles (Parikh, 2017a). Beyond the entrainment velocity, pressure drop starts decreasing and reaches a minimum at terminal air velocity. Settling air velocity, minimum air velocity and terminal air velocity depend upon the physical properties of the particles (e.g. particle size and shape, density and the quantity of the powder in the bowl). A uniform fluidization pattern is essential to obtain the particles of the narrow size distribution. Erratic or no fluidization in the bowl is due to the following reasons.

(a) The large difference in the particle size or the density of the different components of the powder blend.
(b) The powder blend quantity in the bowl is significantly less.
(c) Using of air distribution plate with a low open area may not allow sufficient airflow for the fluidization of the particles.
(d) Extremely cohesive powder bed.

4.3 Components and Functionality of the Fluid Bed Processor

Understanding the different components and their role in the unit operation is very essential for the process optimization and production of the uniform product and scale-up (Pazhayattil et al., 2018). The following section describes different components of the fluid bed processor and its role in process optimization and scale-up and Fig. 4.1 represents different parts of fluid bed processor (Fig. 4.1).

4.3.1 Air Handling Unit (AHU)

The major role of the air handling unit (AHU) is to control the temperature and humidity of the air which is used by the equipment to accomplish different unit operations. It also filters the air using HEPA (High-Efficiency Particulate Arresting) filter before it is used in the processing of pharmaceutical materials.

The drying capacity and hence the evaporation rate depend upon the vapour carrying capacity of the air. A constant drying capacity of the process air is desirable for batch-to-batch uniformity in the process and scale-up (Parikh, 2017a). AHU produces the air which can be used for the processing of the material to obtain the product of desirable quality. Dehumidification of the air is accomplished by passing it over the cooling coils which allow the condensation of the moisture present in the air by reducing the temperature. After processing by AHU, air passes through the air distribution plate to the product container to accomplish the process.

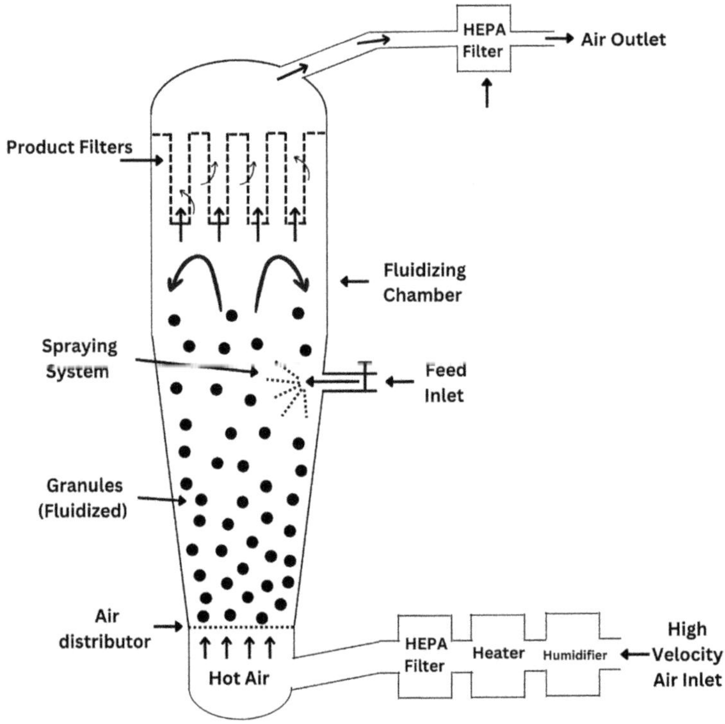

Fig. 4.1 Fluidized bed processor

4.3.2 *Product Container and Air Distribution Plate*

In a fluid bed processor, the product container is of conical shape which is used to hold the material during the process. It has a ratio of 2:1 in the diameter of the upper and lower portion of the conical shape container. At the bottom of the container, an air distribution plate is present to control the airflow rate. The air distribution plates are available in different percentages of open area which varies from 2% to 30%. Since the airflow rate decides the evaporation rate during the drying of granules, the selection of an appropriate air distribution plate is essential to obtain the granules of the desired characteristics. Moreover, the selection of the air distribution plate also depends upon the physical properties (i.e. particle size distribution and density of the material to be processed) (Parikh, 2017b).

4.3.3 *Spray Nozzle*

The spray nozzle is used to deliver the binder/coating solution for the processing of the material in the product container. In the fluid bed processor, a binary nozzle system is usually used in which solvent (one fluid) passes through the orifice and is converted into droplets by compressed air (second fluid). The position of the nozzle in the container depends upon the process to be accomplished. For instance, the top spray is used for the granulation whereas the bottom spray is used for the coating. The number of nozzles to be used depends upon the batch size, for example, for the processing of laboratory-scale batches, a single pot nozzle is sufficient while three- or six-head nozzles are used for the production scale batches to reduce the production time. A peristaltic pump is used to deliver the binder/coating solution to the spray nozzle.

4.3.4 *Filter Bags*

Filter bags made of synthetic polymers (e.g. polyester or nylon) are present at the top of the fluid bed processor. The role of the filter bag is to hold the fine particles which are escaped from the fluidizing bed due to the entrainment air velocity. It is shaken at a specific time interval during the processing to return the fines collected in the bag to the powder bed. The shaking of the filter bag may interfere with the process, as it requires to stop the process for a certain period. However, split filter bags are also available to avoid the above limitation.

4.3.5 *Control Panel*

In the fluid bed processor, a computer-assisted control panel is present to control the process parameter during the granulation or coating process. It enables uniform process conditions throughout the batches and records the process condition for future use.

4.4 Factors Affecting the Granulation Process

Factors that affect the granulation process can be divided into the following three different categories.

4.4.1 Formulation-Related Factors

Several formulation-related factors, including physical properties of the materials, binder and binder solvent, and drug content, play a critical role in the particle size distribution (PSD) and strength of the granules. Physical properties of the materials (e.g. particle size and shape of the particles, density and surface properties) exert a tremendous impact on the granule properties. Uniform particle size with narrow particle size distribution and spherical shape of blend enables the uniform fluidization pattern which generates granules of uniform size with narrow PSD. Spheres exhibit minimum surface area and therefore minimum cohesive/adhesive forces, leading to minimum interaction between the particles during the fluidization which may help in obtaining the granules of the desired size. The presence of large aggregates during fluidization tends to the formation of large granules.

Binder and binder solvent also play an important role in the size and strength of the granules. The selection of binder and binder solvent is a critical factor in the formulation development using the fluid bed processor. For instance, granulation of the formula, which is comprised of high percentage of microcrystalline cellulose with aqueous binder solvent, may result in the formation of small and friable granules. On the other hand, lactose-rich formula if granulated using an aqueous solution of polyvinyl pyrrolidone may produce larger granules. Although the dose of the drug does not produce any direct impact on the granule properties, blend uniformity has been observed as a serious concern. In the case of low-dose drugs, dissolving the drug in the binder solution improves the drug content in fluid bed processes (Levin, 2001).

4.4.2 Process-Related Factors

The process related parameters which affect the fluid bed granulation process are listed below

(a) Inlet air temperature
(b) Air velocity and airflow rate
(c) Spray rate and atomization air pressure
(d) Gun-to-bed distance
(e) Product and exhaust temperature

Process-related parameters affecting the granulation process in fluid bed processors are interdependent and change in one parameter requires an alteration in another parameter simultaneously to obtain the desired product (Lachman et al., 1987). For instance, an increase in airflow rate should be compensated by a proportional increase in the spray rate to maintain the uniformity of the process.

Inlet air temperature is decided based on the solvent used in the granulation process. In the case of aqueous solvent, inlet air temperature is maintained at approximately 40–45 °C, while the 30–32 °C temperature is used if the solvent used for the granulation is organic. Higher temperature may produce smaller and friable granules while lower inlet temperature produce larger and strong granules, and there is a possibility of lump formation due to over-wetting of bed.

Optimization of air velocity and airflow rate is also very critical to obtaining the desired granules (Carstensen, 2000). Airflow rate decides the evaporation rate of the moisture and air velocity decides the fluidization pattern. Higher airflow rate results in increase in drying capacity of the air which leads to an increase in the evaporation rate of the solvent, and the granules produced are small. Higher air velocity increases the fluidized bed height, leading to the formation of fines due to the increase in attrition.

Spray rate also exerts a profound impact on the granule properties. It is always desirable to use the maximum spray rate to keep the production time minimum. However, a higher spray rate may lead to the formation of larger granules due to the large droplet size. Droplet size depends upon the ratio of the mass of the liquid to air in the droplet. To obtain the droplets of the desired size, atomization air pressure also is increased with an increase in the spray rate. Small droplet size may increase the chances of spray drying of the solvent before it reaches the fluidizing bed. It can be avoided by decreasing the distance between gun-to-bed (Stenlake, 1968).

Product temperature decides the rate of evaporation of the solvent. Higher product temperature may produce smaller granules, which could be due to the absorption of the binder solvent from the surface of the particles before substantial particle growth. Exhaust air temperature provides an idea about the temperature and humidity condition in the product container. An increase in exhaust air temperature indicates an increase in the product temperature. Control over the exhaust air temperature can be used as an indirect method to control the product temperature.

4.4.3 Equipment-Related Factors

Equipment design and geometry play a very critical role in the granulation and drying process. Fluid bed processors are conical in shape and the diameter of the upper portion of the equipment is double that of the diameter of the lower portion. The conical shape of the equipment results in a decrease in the air velocity as it moves in the upward direction in the equipment. The lower velocity at the upper part of the equipment is required to prevent the entrainment of the particles into the filter bag. Moreover, the design of the fluid bed processor plays an important role during the scale-up of the granulation process. The airflow rate in the equipment decides the evaporation rate of the solvent during the drying and it should be proportional to the batch size (Gavi & Dischinger, 2021). It has been seen that in laboratory-scale equipment, airflow rate is significantly high as compared to the production scale which results in higher production time. Therefore, a proportional increase in the airflow rate across the scales is desirable in the equipment to obtain the granules of desired properties.

4.5 Process Scale-Up

Granulation using a fluid bed processor is a complex process and requires an understanding of the equipment functionality and process-related variables to obtain a product of similar specifications. The following section describes the strategies for the calculation of various process variables. In general, during the scale-up of the fluid bed process, the below protocol can be followed.

(a) Use the maximum airflow rate as recommended by the manufacturer for the unit.
(b) Calculate the spray rate for the larger batch by maintaining the constant drying capacity.
(c) Maintain the air velocity constant across the scale.
(d) Maintain the same droplet size across the scale by adjusting atomization air pressure.
(e) Adjust the inlet air temperature based on the spray rate and airflow rate.

4.6 Scale-Up Principles

4.6.1 Constant Fluidization Velocity

The fluidization pattern of the particles in the equipment is decided by the fluidization air velocity. A similarity in the fluidization air velocity ensures the similar granule properties obtain from the different processes. Fluidization velocity should be equal to the minimum fluidization velocity to maintain the particles suspended in the air. A narrow particle size distribution of the blend is desirable to avoid the entrainment of the small particles. A constant air velocity across the scale results in the maintenance of a similar fluidization pattern across the scale, leading to the formation of granules of the same particle size (Parikh, 2017b). Fluidization velocity can be calculated by using Eq. 4.1

$$\text{Fluidization air velocity } (\text{m/sec}) = \text{airflow rate (CMH)/cross} - \text{sectional area of the air distribution plate } \left(\text{m}^2\right) \times 3600 \tag{4.1}$$

4.6.2 Airflow Rate

The evaporation rate of the solvent during the drying of granules depends upon the airflow rate. A higher airflow rate presents a high drying capacity of the air. In a laboratory-size fluid bed processor, the airflow rate is high which provides high evaporation as compared to the larger equipment. Generally, it has been observed that the airflow rate is not increased proportionally to the size of the fluid bed processor, leading to higher drying time in pilot or production batches compared to R&D batches. Higher drying time may produce more fines due to more attrition in large batches. During the scale-up, the airflow rate in the pilot/production fluid bed processor can be calculated according to the cross-sectional area of the air distribution plate of the equipment (Edward, 2013). An increase in the airflow rate is directly proportional to an increase in the cross-sectional area of the equipment. Therefore,

$$\frac{A1}{A2} = \frac{V1}{V2} \tag{4.2}$$

Where, $V1$ = airflow rate in small equipment

$A1$ = cross-sectional area of the small equipment
$V2$ = airflow rate in large equipment
$A2$ = cross-sectional area of the large equipment

4.6.3 *Spray Rate*

Spray rate during the granulation process affects the drying capacity of the air. To maintain similar granule properties, drying capacity of the air should be kept constant across the scale. Nevertheless, an increase in the spray rate is desirable to reduce the product time. Total spray rate is increased in large-scale batches by increasing the number of spray nozzles and increasing the spray rate from one nozzle. During the batch size increase in the granulation by fluid bed processor, spray rate is increased proportional to the airflow rate to maintain the drying capacity constant across the scale (Levin, 2001). Spray rate in the large-scale equipment is calculated using the following equation.

$$\frac{SR1}{SR2} = \frac{V1}{V2} \tag{4.3}$$

Where, $V1$ = airflow rate in small equipment

$SR1$ = spray rate of the small equipment
$V2$ = airflow rate in large equipment
$SR2$ = spray rate of the large equipment

4.6.4 *Droplet Size*

Droplet size exerts a significant impact on the particle size of the granules. Large binder droplet produces larger granules as compared to the smaller ones. Therefore, to maintain the granule properties similar across the scale, droplet size should be kept constant. Droplet size is decided by the ratio of the mass of air to the mass of liquid.

An increase in spray rate is desirable in large-scale batches to reduce the production time. However, an increase in the spray rate should be compensated by an increase in the atomization air pressure to maintain the same droplet size. However, depending upon the ability of the equipment to atomize the air, it may not be possible to go beyond certain atomization air pressure. In that case, the spray rate should be reduced to maintain a similar droplet size and the reduced spray rate is compensated by changing the inlet air temperature to keep the drying capacity constant (Levin, 2001). For example, during the increase in a batch size of a fluid bed granulation process, it is decided to increase the airflow rate by 10 times (based on the cross-sectional area of the fluid bed processor). According to the scale-up principle, the spray rate should be increased proportionally to the airflow rate. However, an 8 times increase in the spray rate is possible in the equipment. This difference can be compensated by an increase in atomization air pressure and a decrease in inlet air temperature.

Example 4.1 A batch size of the fluid bed granulation process was increased from GPCG 60 to GPCG 300. The cross-sectional area of the fluid bed processer is 0.416 m^2 and 1.038 m^2, respectively. A 300 cfm of airflow rate and 400 gm/min spray rate were used in GPCG 60 to obtain the granules of desired properties. Calculate the fluidization air velocity, airflow rate and spray rate for the GPCG 300.

(a) *Calculation of fluidization air velocity*

Fluidization air velocity can be calculated by using Eq. 4.1

$$\text{Fluidization air velocity} = \text{airflow rate}/\text{cross-sectional area}$$
$$= 0.13/0.416 = 0.31\text{m/s} \ (300 \text{ cfm} = 0.13 \text{ CMH})$$

Fluidization air velocity should be kept constant across the scale to maintain a similar fluidization pattern. Therefore, fluidization air velocity for the GPCG 300 is 0.31 m/s.

(b) *Calculation of airflow rate for the GPCG 300*

To maintain the constant evaporation rate across the scale, the airflow rate should be increased proportionally to the cross-sectional area of the equipment.

$$\frac{A1}{A2} = \frac{V1}{V2}$$

$$\text{Airflow rate in GPCG 300} = (A2/A1) \times V1$$
$$= (1.038/0.416) \times 509.7 = 1271.7 \text{ CMH} = 749 \text{ cfm}$$

Airflow rate can also be calculated by using the fluidization air velocity. (i.e. fluidization air velocity should be kept constant across the scale). Therefore, the airflow rate in GPCG 300 is

$$
\begin{aligned}
&= \text{fluidization air velocity} \times \text{cross-sectional area} \\
&= 0.31 \times 1.038 \\
&= 0.32\ \text{m}^3/\text{s} = 1159\ \text{CMH}
\end{aligned}
$$

(c) *Calculation of spray rate for the GPCG 300*

As the size of the equipment increases, the drying capacity of air in large-scale equipment also increases, leading to an increase in the evaporation rate of the solvent. Therefore, drying capacity must be kept constant across the scale to obtain granules of similar properties. Therefore, the spray rate for the GPCG 300 is calculated using the following equation.

$$\frac{SR1}{SR2} = \frac{V1}{V2}$$

$$SR2 = (V2/V1) \times SR1 = (749/300) \times 400 = 998\ \text{gm}/\min$$

References

Alderborn, G., & Aulton, M. E. (2002). *Pharmaceutics: The science of dosage form design*. Harcourt Publishers Limited.

Carstensen, J. T. (2000). Advanced pharmaceutical solids. *Advanced Pharmaceutical Solids, 110*, 1–510. https://doi.org/10.1201/b16941

Carter, S. J. (1996). *Cooper and Gunn tutorial-pharmacy*. CBS Publishers.

Edward, G. J. (2013). *Bring fluid bed granulation up to scale | Pharma manufacturing, pharmaceutical manufacturing magazine*. Available at https://www.pharmamanufacturing.com/home/article/11326986/bring-fluid-bed-granulation-up-to-scale. Accessed 8 Sept 2022.

Gavi, E., & Dischinger, A. (2021). Scale-up of fluid bed granulation using a scale-independent parameter and a process model. *AAPS PharmSciTech, 22*(4), 1–10. https://doi.org/10.1208/S12249-021-02013-X/TABLES/4

Lachman, L., Liberman, H. A., & Kanig, J. L. (1987). *The theory and practice of industrial pharmacy*. Varghese Publishing House.

Levin, M. (Ed.). (2001). Pharmaceutical process scale-up. CRC Press.

Parikh, D. M. (2017a). *How to optimize fluid bed processing technology*. Academic/Elsevier.

Parikh, D. M. (2017b). Process scale-up. In *How to optimize fluid bed processing technology* (pp. 127–150). https://doi.org/10.1016/b978-0-12-804727-9.00011-9

Pazhayattil, A. B., et al. (2018). *Solid oral dose process validation*. AAPS Introductions in the Pharmaceutical Sciences.

Stenlake, J. B. (1968). *Pharmaceutical monographs*. William Heinemann Medical Books Ltd.

Chapter 5
Drying

Anil B. Jindal

Contents

Abstract Drying is a vital process in pharmaceutical formulation involving removal of moisture from solids by applying heat, thus resulting in both heat and mass transfer. The process when theoretically studied utilizes various terms such as absolute humidity, saturation humidity, dew point, relative humidity, drying capacity, wet bulb temperature and the dry bulb temperature which can be understood with the help of psychometric chart. The process of drying of solids when graphically plotted as drying rate versus moisture content is known as drying rate curve. The drying rate curve can be classified as initial adjustment period, constant rate period, first falling rate period and second falling rate period as per the trends of curve. Various types of equipment are used for drying the pharmaceutical solids. The most basic amongst them is a tray drier where the solids are kept on a tray and are dried with hot air. Whereas in fluidized bed dryer, the solids are dried by suspending in air by a constant stream of hot air. On the other hand, when the fluid content is higher, spray drying can be used where the mixture is pumped on a hot drying environment leading to atomization and further leaving particulate solid residue of the mixture.

A. B. Jindal (✉)
Department of Pharmacy, Birla Institute of Technology and Science, Pilani, Jhunjhunu, Rajasthan, India
e-mail: anil.jindal@pilani.bits-pilani.ac.in

A. B. Jindal (ed.), *Pharmaceutical Process Engineering and Scale-up Principles*, AAPS Introductions in the Pharmaceutical Sciences 13,
https://doi.org/10.1007/978-3-031-31380-6_5

Freeze drying and vacuum drying are the specialized drying processes where drying is achieved by freezing samples or by reducing the pressure around the samples, thus resulting in drying without heating.

Keywords Drying · Drying rate curve · Freeze drying · Spray drying

5.1 Introduction

In the pharmaceutical industry, drying is an important unit operation in the manufacturing of tablet and capsule dosage forms. In most cases, the manufacturing of tablets includes a wet granulation process. Granules of desired particle size and size distribution are obtained by using aqueous or organic solvents and a binding agent. The solvent is removed from the wet granules by drying process using a suitable dryer (Carter, 1996; Subramhanyam, 2001). Drying is also used to modify the physical properties of the excipients to make it more suitable for processing in tablets and capsules. For example, spray drying of lactose improves the flow properties of the excipient during the manufacturing of solid dosage forms.

Drying is defined as the removal of water from the solids to the surrounding unsaturated air by applying heat. It involves both heat and mass transfer phenomena (Alderborn & Aulton, 2002). Nonthermal methods of drying, such as expression, extraction and adsorption, are not discussed in this chapter. The chapter focuses on the thermal method of drying pharmaceutical solids using different dryers.

5.2 Psychometry

The rate and extent of drying are dependent upon the drying capacity of the air sued for the drying purpose. The drying capacity of the air can be defined as the ability of the air to hold the amount of water vapour at constant temperature and pressure. An increase in the temperature of the air leads to an increase in drying capacity due to an increase in unsaturation (Lachman et al., 1987). Studying the air-water vapour system at constant pressure with a drying capacity of the air is termed psychometry (Peralta, 2005). Psychometry is used for the determination of the operating condition of drying and plays a very important role in drying unit operation. The psychometric chart represents the relationship between air and absolute humidity at constant pressure and is used to study the air-water system. A standard psychometric chart is presented in Fig. 5.1.

The psychometric chart has been discussed here with respect to the following key term: absolute humidity, dry bulb temperature, wet bulb temperature, saturation humidity and dew point. The concentration of water vapour in the air is known as humidity absolute humidity is defined as the amount of water vapour present in the

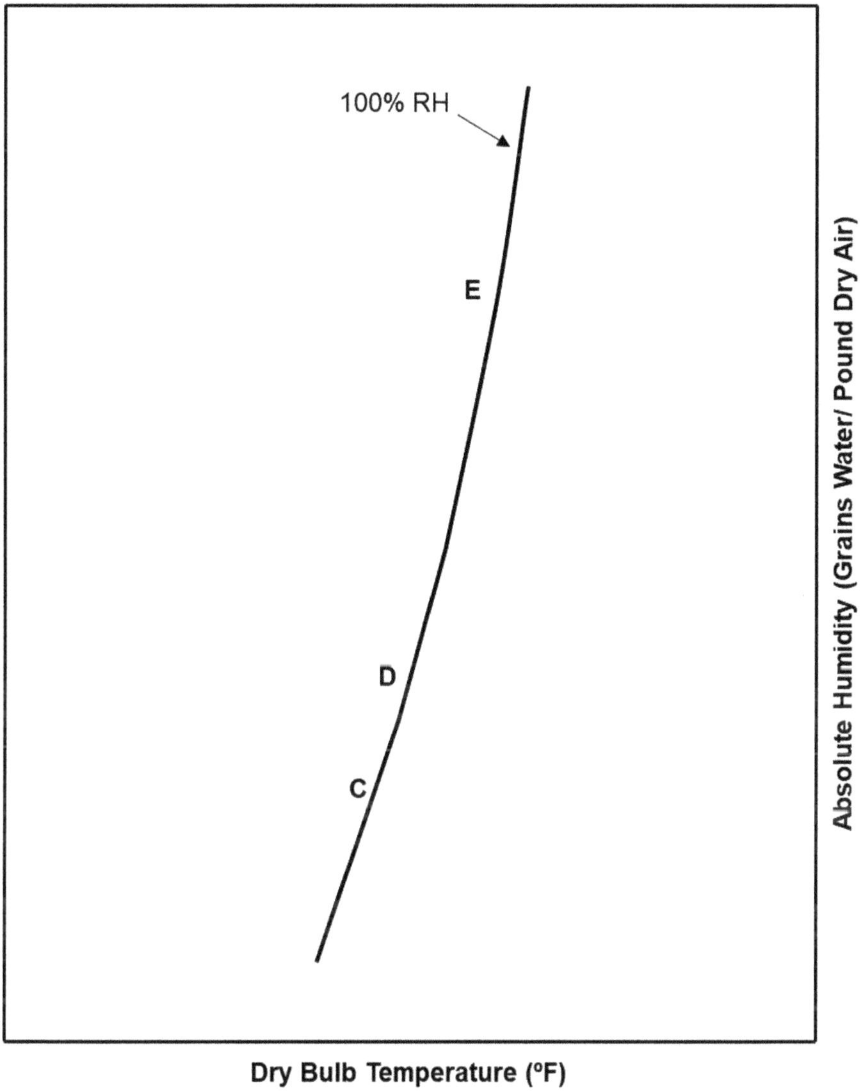

Fig. 5.1 Psychometric chart

unit weight of the air. Saturation humidity is defined as the absolute humidity when the partial pressure of the water vapour is equal to the vapour pressure of the free water. In the psychometric chart shown in Fig. 5.1, curve CDE represents the saturation humidity curve. In saturated humidity conditions, the air is completely saturated with moisture and cannot be used for drying purposes.

At the saturation humidity curve, point C is referred to as a dew point which can be defined as the temperature at which air can hold the maximum amount of moisture without condensation. Condensation of moisture takes place if the temperature of the air is reduced below the dew point. At the dew point, the air is saturated with water vapour and cannot be used for drying purposes (Hosahalli S. Ramaswamy, 2005). The humidity of the air varies from season to season during the year and location of the plant (e.g. it is significantly high in coastal areas). Dew point can be used to define the drying temperature of the heated air in different humidity conditions. A temperature higher than the dew point leads to unsaturation of the air which can be used for drying.

The percent relative humidity indicates the humidity of the air about the saturation. It can be defined as the ratio of absolute humidity of the air to the ratio of absolute humidity of the saturated air. Percent relative humidity can also be considered as a measure of the drying capacity of the air. Higher the drying capacity, the higher the rate and extent of drying.

To understand the wet-bulb temperature, it is important to understand the process of mass transfer and heat transfer during drying. Evaporation of the moisture from the surface of the solids takes place due to the difference between the vapour pressure of the surface water and the surrounding air which results in the decrease of the surface temperature due to latent heat of vaporization. It ultimately leads to the development of the temperature gradient between the surface and surrounding air, which leads to an increase in the rate of heat transfer. After a certain period, the rate of heat transfer becomes equal to the rate of evaporation and an equilibrium is established which results in the stabilization of the surface temperature and is referred to as wet-bulb temperature (Lachman et al., 1987). The rate of drying becomes constant at the wet-bulb temperature. Dry bulb temperature is defined as the temperature which is determined by an ordinary thermometer.

5.3 Drying Cycle

The drying process involves both mass and heat transfer. Heat transfer involves the transfer of heat to the wet surface through the heated air for the evaporation of the liquid present on the surface. While mass transfer is involved in the transfer of the liquid from the internal structure of the material to the surface, and from the surface to the surrounding environment. The driving force for the mass transfer is the difference between vapour pressure and partial pressure for the water present on the surface and the temperature gradient is the driving force for the heat transfer operation (Carter, 1996). A typical drying curve plotted moisture content vs drying rate, obtained from the drying of granules is presented in the Fig. 5.2.

Atypical drying cycle can be divided into five different phases namely: (a) initial adjustment phase, (b) constant rate period, (c) First falling rate period, (d) second falling rate period and (e) equilibrium moisture content.

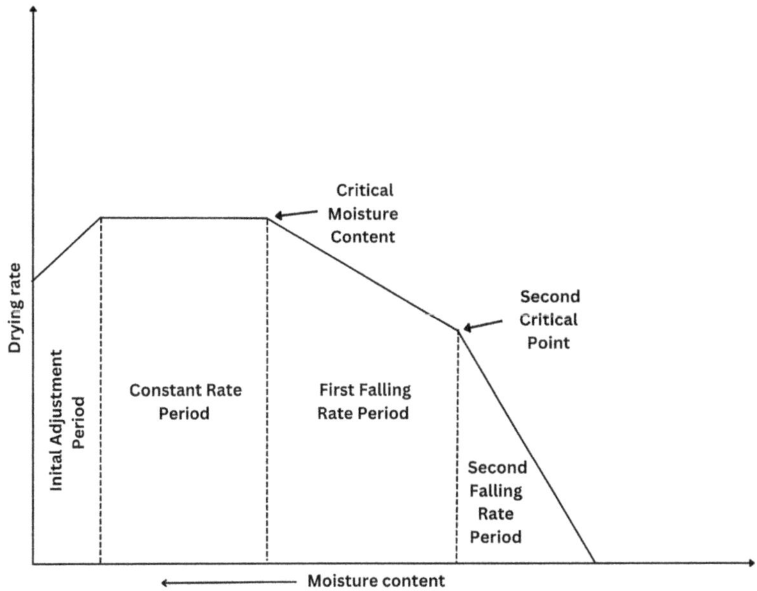

Fig. 5.2 Atypical drying curve

5.3.1 *Initial Adjustment Period*

During the initial adjustment phase, when a significant amount of moisture is present on the surface, heat is transferred from the heated air to the wet surface which results in the evaporation of the liquid from the surface due to the latent heat of vaporization. Evaporation of the surface liquid leads to a decrease in the surface temperature which establishes a temperature gradient for the heat transfer process. The initial adjustment phase continues until an equilibrium is established between mass transfer and heat transfer processes.

5.3.2 *Constant Rate Period*

When equilibrium is established between heat transfer and mass transfer processes, the temperature of the surface reaches the wet-bulb temperature and remains constant. During the constant rate period, a continuous film of the liquid is always present on the surface. During this period, liquid evaporated from the surface is quickly replaced by the moisture diffused from the material's internal structure. During a constant rate period, the rate of diffusion of the liquid from the internal structure is equal to the rate of evaporation of the liquid from the surface and the drying rate remains constant. The moisture content which is present in the material at the end of the constant rate period is known as critical moisture content.

5.3.3 First Falling Rate Period

When the rate of diffusion of the moisture from the internal structure of the material to the surface becomes slower than the rate of evaporation of the moisture, a dry spot starts appearing on the material surface. The drying rate during the first falling rate period depends on the rate of diffusion of the moisture to the surface and it starts decreasing. Internal mass transfer during this phase is defined by the internal structure of the material, i.e. presence of capillaries or pores in the material.

5.3.4 Second Falling Rate Period

During the second falling rate period, the drying rate further decreases and the process of removal of water becomes slow. There is no moisture present on the surface during this period. Moisture present inside the material reaches the surface by diffusion and subsequently evaporates from the surface to the surrounding heated air. The second falling rate period is completed when the moisture of the material reaches the equilibrium moisture content.

5.3.5 Equilibrium Moisture Content

When the drying rate becomes zero, the moisture present in the material is known as equilibrium moisture content. The amount of moisture present in the material at this stage cannot be removed due to the saturation of surrounding air. To decrease the equilibrium moisture content of the sample, the temperature of the air should be increased to make it unsaturated.

5.4 Drying Equipment

5.4.1 Tray Dryer

Tray dryer is commonly used for the drying of granules in the pharmaceutical industry. However, it has been replaced by modern drying equipment available for drying due to the inefficient and slow drying process in the tray dryer. The design of the tray dryer is very simple which consists of a cabinet and a few metal trays to hold the granules during drying (Fig. 5.3). The number of trays varies from three to twenty depending on the size of the equipment. The metal tray can be solid or perforated depending on the drying purpose. It has a very convenient drying operation. The wet granules are spread like a powder bed onto the metal tray inside the dryer and heated air is passed through the granules. The drying of granules is accomplished by the transfer of heat from the circulated air by the convection mechanism. Heated air is circulated in the dryer by the fans mounted on the sidewalls of the dryer. As the heated air passes through the bed of granules, moisture diffuses from inside the bed towards the top which is eventually evaporated from the surface. The drying rate largely depends on the thickness of the bed of the granules onto the tray and an increase in the bed thickness increases the drying time due to an increase in the diffusional path length of the water (Subramhanyam, 2001). The air is heated by using steam or electricity. However, steam is preferred due to the cheap source of energy. Drying using tray dryers is a very slow process and may take 48 h to dry one batch of granules. Another major limitation in the use of tray dryers is that it is a batch process and involves lots of labour costs due to the requirement of repeated loading and unloading of the dryer. Moreover, there is a chance of moving the dye or any other soluble material on the surface of the granules during the drying operation in the tray dryer. A typical drying cycle of the granules dried using tray dryers can be characterized by a short constant rate period and a large falling rate period. The large falling rate period can be explained by the fact that the drying of granules in the tray dryer depends on the diffusion of the water from inside the bed towards the surface.

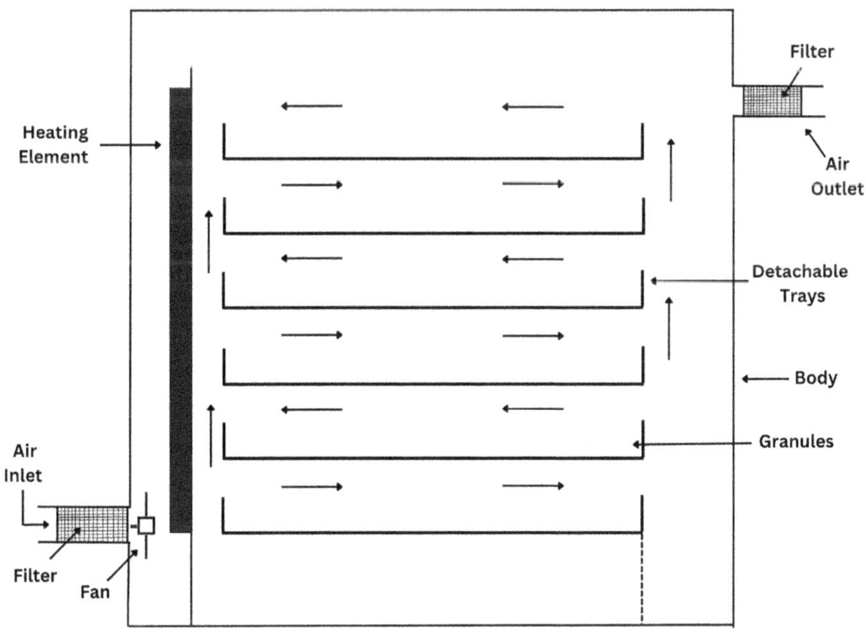

Fig. 5.3 Tray dryer

The diffusion of the moisture inside the bed takes place through the void space of the granules. Larger the void space, the faster the rate of drying which may lead to a reduction in the drying cycle although it may not contribute significantly to the reduction of total production time due to the considerably large drying time. The equilibrium moisture content of the granules dried in the tray dryers depends on the temperature of the heated air circuited for drying. To decrease the equilibrium moisture content of the dried granules, it is necessary to increase the temperature of the air to avoid saturation. Nevertheless, while increasing the temperature of the air, the impact of high temperature on the stability of the active pharmaceutical ingredients should be considered (Lachman et al., 1987).

5.4.2 *Fluidized Bed Dryer*

The major limitation of the tray dryer is the slow and inefficient drying process which is due to the limited surface area exposed to the heated air. In a fluidized bed dryer, granules remain fluidized in the heated air which allows quick drying due to the exposure of wet granules to the heated air in all directions. A batch fluidized bed dryer consists of a fan to provide fluidized air, a drying chamber and a filter beg which is present at the top of the dryer (Fig. 5.4). An air distribution plate is present at the bottom of the dryer to support the granules. The air flows in the dryer in the upward direction through the bed of granules supported by an air distribution plate. At low air velocity, the granule bed is static which starts fluidizing with an increase in the air velocity. The air velocity should be maintained above the incipient velocity and below the entrainment velocity of the granules. Fine particles which are escaped from the fluidization zone are collected in the filter bag on the top of the dryer and can be added back to the granules by shaking the bag during drying. A typical drying time in a fluidized bed dryer varies between 15 and 45 min depending on the solvent used for drying. The drying rate depends on the inlet air temperature and airflow rate. An increase in both inlet air temperature and airflow rate may result in faster drying of the granules, probably due to an increase in both heat and mass transfer processes. The particle size distribution and strength of the dried granules depend on the inlet air temperature and binder solvent. In the case of organic binder solvent, the higher inlet air temperature may result in friable and small granules. At a high airflow rate, particles present in the per unit volume of the air decrease, leading to an increase in drying rate due to an increase in the surface area exposed to the heated air. A typical drying cycle in a fluidized bed dryer exhibits a long constant rate period followed by a short falling rate period. Due to the fluidization, granules are exposed to the heated air from all directions which enables drying at a constant rate until evaporation of the maximum amount of moisture present in the granule. In the case of aqueous solvents, the drying rate largely depends on the humidity of the air, as the driving force for the mass transfer process is the difference in the vapour pressure of the moisture present in the air and wet solid. During the drying process, moisture evaporates from the wet granules and enters the surrounding unsaturated air which

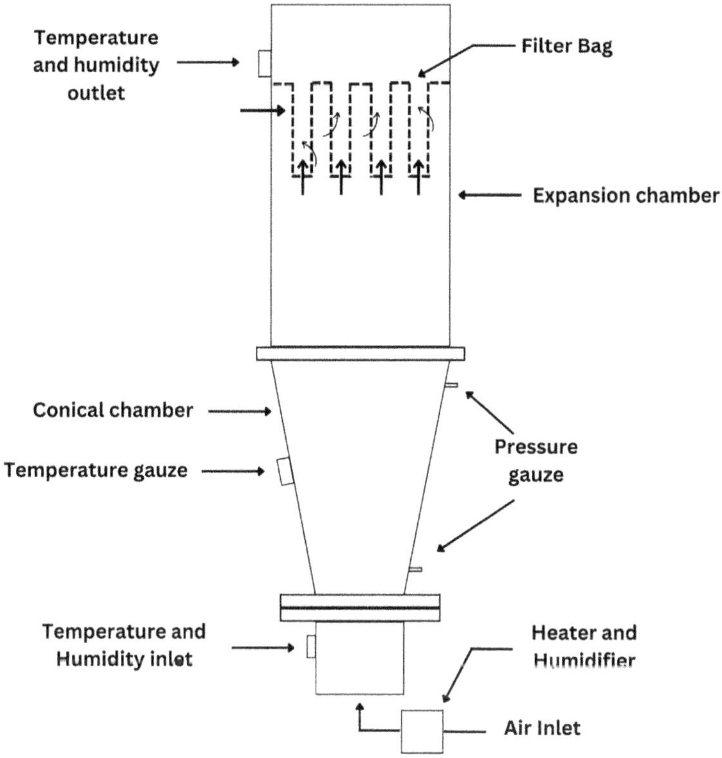

Fig. 5.4 Fluidized bed dryer

results in a slight decrease in the air temperature. A constant difference between the inlet and exhaust air temperature indicates the drying at a constant rate. When the process enters the falling rate period, an increase in the exhaust air temperature is observed due to the decrease in the evaporation rate of the moisture, leading to an increase in the product temperature. It is therefore essential to decrease the inlet air temperature during the falling rate period to avoid overheating of the granules.

Fluidized bed dryers are most widely used in the industry for the drying of granules wherein the fluidization of the granules results in a decrease in the drying time from days to a few minutes. The significant decrease in the drying time results in the overall decrease in the production time, which makes the process cost-effective (Parikh, 2017).

5.4.3 Spray Dryer

A spray dryer is used for the drying of materials in the form of solutions or slurries. It is significantly different from the other dryers in the drying purpose. The process of spray drying involves converting the feed into fine droplets by using the spray nozzle. The liquid present in the droplet evaporates when it comes to contact with the heated air circulating in the dryer. After the evaporation of the liquid, the dried material is separated from the airstream in the collection system. When a liquid droplet comes in contact with the heated air, the liquid present on the droplet surface quickly evaporates due to the high surface area. It leads to the formation of a thick layer of the solid encapsulating the remaining liquid inside it. As the material is still in contact with the heated air, heat is transferred through the solid layer from inside the droplet, leading to evaporation of the liquid (Padma Ishwarya & Anandharamakrishnan, 2017). The rate of heat transfer is less than the rate of diffusion of the solvent vapour from inside the droplet to the surface through the solid layer present on the surface which results in building up pressure within the droplet (Anandharamakrishnan & Ishwarya, 2015). It tends to rupture the outer solid layer to release the internal pressure (Principle, Construction, Working, Uses, Merits and Demerits of Spray Dryer: Pharmaguideline, 2007). It is therefore spray-dried material consisting of ruptured spheres.

A typical spray dryer consists of three basic components, (1) Feed delivery system, (2) spray nozzle and (3) cyclone separator (Fig. 5.5). A feed delivery system pumps the liquid at a constant rate through the spray nozzle into the drying chamber (*Spray Drying - Freund Vector*, 2013). The role of the spray nozzle is to convert the feed material into small droplets by atomization to enable quick drying. The spray nozzles accomplish the atomization processes by using high pressure or impacting the feed droplet onto the disc rotating at high speed. The dried material is separated from the airstream by the gravitational and centrifugal forces.

It is interesting to note that although the spray drying process involves the use of high temperature for the drying purpose, it is suitable for the drying of thermolabile substances. It is because in the spray dryer liquid droplets are dried very quickly usually in a fraction of seconds due to the presence of high surface area in contact with hot air and dried material is removed rapidly from the dryers. Therefore, in the spray dryers, the material remains at a high temperature for a very short period which does not produce any effect on the stability of the product (Pazhayattil et al., 2018).

Several pharmaceutical excipients are also available as spray-dried grades (e.g. spray-dried lactose for use in the manufacturing of solid oral dosage forms). Spray-dried materials exhibit unique shape and density, which is the result of the drying mechanism used in spray dryers. The spherical particle shape of the spray-dried product improves the flow properties of the material by offering low frictional forces between the particles due to the minimum surface area.

Spray drying is also used for the encapsulation of solids and liquids for different purposes. For instance, to control the drug release, to mask the taste or to improve the physical stability of the product. Encapsulation of solids by spray drying is

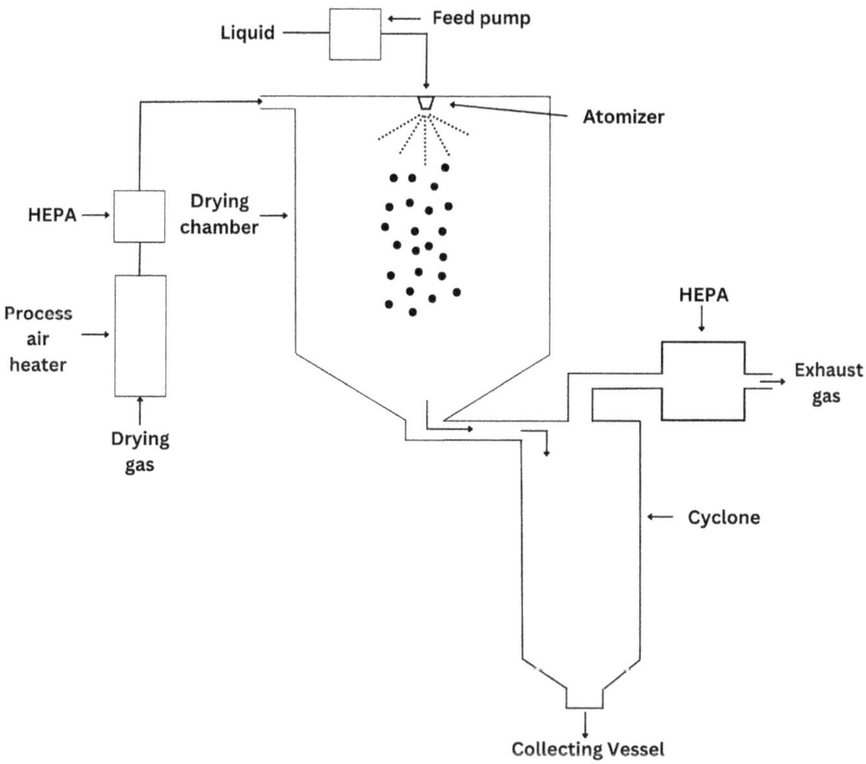

Fig. 5.5 Spray dryer

accomplished by dispersing the solid particles into a coating solution which is subsequently converted into small droplets and spray-dried into fine particles. It is mainly used for the sustained release of the drug. Liquid material is encapsulated by preparing an emulsion of the core liquid material into an aqueous coating solution. For example, an aqueous solution of gums in water. The resulting emulsion is converted into small droplets by atomization and spray-dried using hot air. It is mainly used for the encapsulation of volatile oil to increase the shelf life.

5.5 Specialized Drying Techniques

5.5.1 Freeze Drying

Several pharmaceutical products are available in the form of powder for reconstitution intended for administration by ophthalmic, parenteral, oral, and nasal routes. The dried products are generally redissolved in the appropriate grade of water,

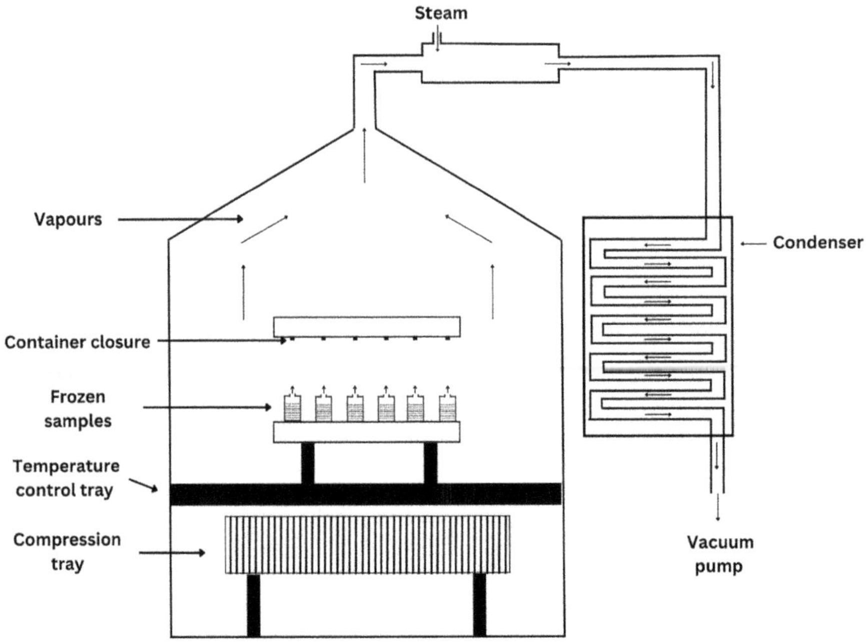

Fig. 5.6 Freeze dryer

provided by the manufacturer in the drug product package, just before use. Several antibiotics are also formulated in the form of powder for reconstitution for paediatric use. Those drugs which are either heat sensitive or not stable in presence of oxygen are usually formulated in the form of powder for reconstitution. The drying technique which is used for the manufacturing of such products is freeze drying and is also referred to as lyophilization. It involves the freezing of the material which is to be dried followed by sublimation of the frozen water under high vacuum and low temperature to obtain the solid product. Freeze-dried products are amorphous in nature and tend to absorb moisture from the surrounding environment.

Freeze drying uses the principle of sublimation wherein water in a solid-state (ice) is directly converted into a gaseous state (vapour) without forming a liquid (Nail et al., 2002). Sublimation of water happens at a specific temperature and pressure conditions i.e. temperature and pressure below the triple point of the water. In freeze drying, it is necessary to maintain a positive vapour pressure on the surface of the solids, i.e. vapour pressure of water in the solids should be higher than the partial pressure of the water in the vapour and moisture should be continuously removed from the vapours (Santos et al., 2017). A freeze dryer contains a vacuum chamber that can hold the material during drying, a vacuum pump to apply the vacuum, a heat source and an arrangement to remove the vapours (Fig. 5.6). The detailed construction of a freeze dryer is out of the scope of this chapter and not discussed here.

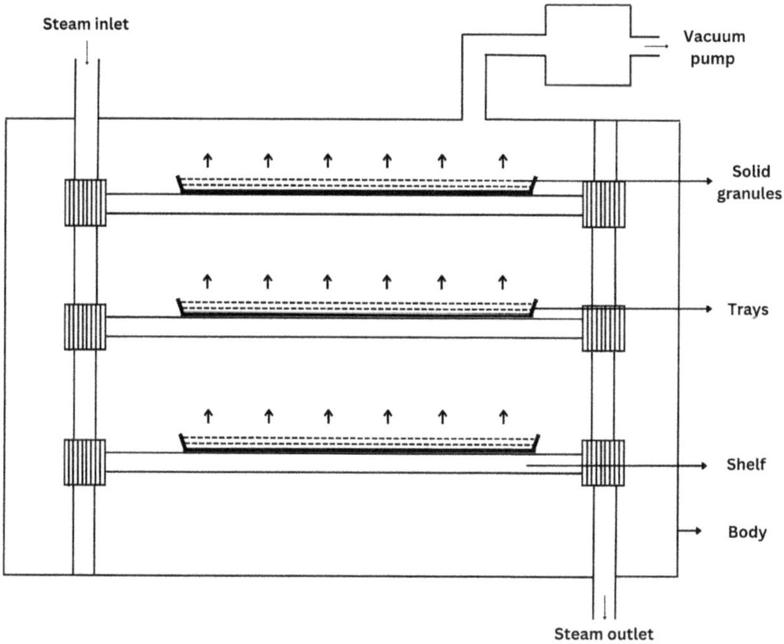

Fig. 5.7 Vacuum dryer

5.5.2 *Vacuum Drying*

The driving force for the drying process is the difference between the vapour pressure of the water in the solids and the partial pressure of water in the surrounding air. It is also directly proportional to the area of the solids exposed to the heated air and inversely proportional to the heat of evaporation. In the case of vacuum drying, the atmospheric pressure in the surroundings of the wet solid is reduced by applying a vacuum using a vacuum pump. It tends to increase the evaporation of the water from the solids at a lower temperature. The main advantage of vacuum drying is that the technique can be used for the drying of heat-sensitive materials because the drying can be accomplished at a low temperature (Lachman et al., 1987). The rate of drying in the case of a vacuum dryer is comparable to a tray dryer (12–48 h) and it is significantly less than freeze drying (Fig. 5.7).

5.6 Conclusion

Drying remains a key process in pharmaceutical formulation development. Psychometric chart and drying curve can be used for the thorough interpretation of the process. Dried product can be obtained with various methods using different instruments. It must be noted that the physicochemical characteristics of the constituents are needed to be understood for the selection of the process of drying. Hence, a thorough knowledge of the process is essential for the user for its proper application.

References

Alderborn, G., & Aulton, M. E. (2002). *Pharmaceutics: The science of dosage form design*. Pharmaceutics: The Science of Dosage form Design.

Anandharamakrishnan, C., & Ishwarya, S. P. (2015). *Introduction to spray drying* (pp. 1–36). Spray Drying Techniques for Food Ingredient Encapsulation. https://doi.org/10.1002/9781118863985.ch1

Carter, S. J. (1996). *Cooper and Gunn tutorial-pharmacy*. CBS Publishers.

Lachman, L., Liberman, H. A., & Kanig, J. L. (1987). *The theory and practice of industrial pharmacy*. Varghese Publishing House.

Nail, S. L., et al. (2002). Fundamentals of freeze-drying. *Pharmaceutical Biotechnology, 14*, 281–360. https://doi.org/10.1007/978-1-4615-0549-5_6

Padma Ishwarya, S., & Anandharamakrishnan, C. (2017). *2 spray drying: Principle of operation* (pp. 58–93). Handbook of Drying for Dairy Products.

Parikh, D. M. (2017). *How to optimize FLuid bed processing technology*. Academic Press, Elsevier.

Pazhayattil, A. B., et al. (2018). *Solid oral dose process validation*. AAPS Introductions in the Pharmaceutical Sciences.

Peralta, P. (2005). *The psychrometric chart: Theory and application* (p. 260). NC State University.

Principle, Construction, Working, Uses, Merits and Demerits of Spray Dryer: Pharmaguideline. (2007). Available at: https://www.pharmaguideline.com/2007/02/principle-construction-working-uses-merits-demerits-of-spray-dryer.html. Accessed 20 Dec 2022

Ramaswamy, H. S., & Marcotte, M. (2005). *Food processing principles and applications* (1st ed.). Taylor & Francis Group. https://doi.org/10.1201/9780203485248

Santos, D., et al. (2017). *Spray drying: An overview*. Biomaterials – Physics and Chemistry – New Edition. https://doi.org/10.5772/INTECHOPEN.72247

Spray Drying – Freund Vector (2013). Available at: https://www.freund-vector.com/technology/spray-drying/. Accessed 8 Jan 2023

Subramhanyam, C. V. S. (2001). *Pharmaceutical engineering – Principles and practices*. Vallabh Publisher.

Chapter 6
Compression

Ketan Patel and Anil B. Jindal

Contents

Abstract Compaction of powder involves compression and consolidation of the solids on the application of pressure. The forces involved in these processes play an important role in the design and development of solid oral dosage forms including tablets, filling of the hard-gelatin capsule, and in the handling of powders. Assessment of the forces on punches, axial forces, radial forces, frictional forces, and ejection forces gives us information regarding the compaction behaviour of the powder. Several mathematical terms to describe the compaction processes are studied such as the heckle plot. Optimizing the parameters that affect these forces would produce a product of the desired quality. In this chapter, we have discussed in detail about compression, consolidation, and forces involved in compaction and their impact on the quality of the product.

Keywords Compression of tablets · Compaction force · Consolidation · Hardness · Brittle fracture

K. Patel
College of Pharmacy and Health Sciences, St. John's University, New York City, NY, USA

A. B. Jindal (✉)
Department of Pharmacy, Birla Institute of Technology and Science, Pilani, Jhunjhunu, Rajasthan, India
e-mail: anil.jindal@pilani.bits-pilani.ac.in

A. B. Jindal (ed.), *Pharmaceutical Process Engineering and Scale-up Principles*, AAPS Introductions in the Pharmaceutical Sciences 13,
https://doi.org/10.1007/978-3-031-31380-6_6

6.1 Introduction

Several manufacturing processes involve the use of some degree of force to convert powder blend into a dosage form (e.g. tablet, capsule, or compact powder). Some specific unit operations (e.g. roller compaction) also involve the use of external force. Understanding the effect of different forces on the powder blend properties is essential for the development of different dosage forms. Compression can be defined as the process of bulk volume reduction due to the rearrangement or deformation of the particles under high loads. When the powder blend is subjected to high compressional forces, the strength of the compressed structure increases due to the interparticle interaction. This process is termed consolidation (Carstensen, 2000). Several physical properties of the powder blend, including particle size, particle size distribution, density, and surface properties, significantly impact the powders' compression and consolidation (Pazhayattil et al., 2018).

6.2 Bulk Volume Reduction Process During Compression

During the tableting, when the powder blend is subjected to external forces, the bulk volume of the blend is reduced due to the rearrangement of the particles depending on the particle size and void space available (Fig. 6.1). The rearrangement of the particle leads to the repacking of the powder blend in the die cavity (Jones, 2008). When there is no space available for the rearrangement, particles start deforming either elastically or plastically. The mechanism of deformation depends upon the amount of force applied and the internal structure of the materials. If particles regain their original state after removing the load, the deformation is termed elastic deformation. Depending upon the applied load, all the materials can undergo elastic deformation. When the applied load is higher than the elastic limit of the material, particles deform plastically (i.e. they do not regain their original state after removal of the load). At a high load, when the shear force working on the particles is greater than their breaking strength, particles undergo brittle fracture and fines generated from it occupy the void space, leading to a reduction in bulk volume. Large and brittle particles undergo brittle fracture more frequently than small particles. The deformation mechanism in the case of small particles is plastic in nature (Alderborn & Aulton, 2002). In the case of plastically deforming materials, rapid loading and unloading may result in the failure of the tablet structure. This may happen at the large-scale production of tablets when the compression machine speed is too high and can be avoided by reducing the machine speed. When the distance between the particles is significantly low, it undergoes consolidation by interparticle interaction. During the particle deformation by plastic or brittle fracture mechanism, new surfaces are generated. It is therefore essential that enough compression force should be available for bonding. Consolidation of particles is significantly affected by the presence of contaminants on the surface of the particles (Lachman et al., 1989). For

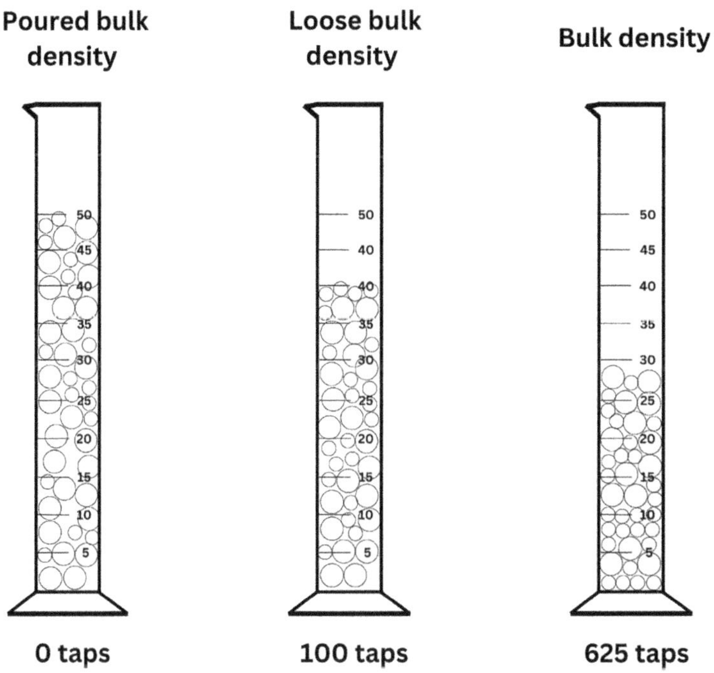

Fig. 6.1 Typical arrangement of powders on the basis of voids

example, over-lubrication of the powder blend may result in a soft tablet, probably due to the poor consolidation strength. Lubricants form weak bonds with the particle surface and due to the absence of direct interaction of the particles poorly consolidated tablet structure formed. The presence of moisture in the powder also affects the consolidation. Due to the presence of high compressional forces, traces of material dissolved in the moisture recrystallize after removal of the compressional forces.

6.3 Compression Under High Load

During the tableting, compression force is high enough to induce plastic deformation and brittle fracture which results in the generation of new surfaces and these new surfaces are consolidated to form a single solid structure (i.e. tablet under high compressional load). The solid structure formed should be strong enough to withstand the stresses generated within the tablet due to the elastic recovery of the materials after the removal of the load.

During the compression of the tablet, a force is applied in the axial direction which results in a decrease in the height of the powder mass in the die cavity. In an unconfined state, a decrease in the height is compensated by an increase in the diameter of the powder mass (Marshall, 1999). However, in the case of tableting, as the powder mass is confined in a die cavity, it is not able to expand in the horizontal direction, leading to the generation of a force in the radial direction. The radial force forms a thick layer of powder mass on the lateral surface of the tablet. This thick layer could prevent penetration of the aqueous medium to the tablet in the GIT and delay the disintegration and dissolution of the tablet. It is therefore desirable to keep the tablet height-to-diameter ratio as minimum as possible (Lachman et al., 1987).

Furthermore, the compression force applied by the upper punch decreases exponentially during the downward movement of the punch in the die cavity. This is because of the presence of frictional forces between the particles and between the die-wall and the particles. It ultimately leads to the non-uniform force distribution in the different regions of the tablet. It can be avoided by decreasing the tablet height-to-diameter ratio and lubrication of the blend.

6.4 Frictional and Radial Forces During Compression

Two types of frictional forces, interparticle friction and die-wall friction forces, are present during powder flow and compression. It hinders the flow of powder from the hopper to the die cavity and the movement of particles during compression. Interparticle frictional forces are effective at low applied compressional forces and the compressional force which is used during the tableting is significantly high, resulting in a negligible impact of interparticle forces on the movement of particles within the die cavity. The predominant effective frictional forces during the compression are due to the die-wall friction which acts in the direction opposite to the upper punch force (Kikuta & Kitamori, 1983).

Another important force that is generated during the compression is the radial force, which acts perpendicular to the die walls. During the compression when minimum porosity is attained at the applied load, powder mass in the die cavity behaves like a solid body. At this stage, when further compressional force is applied, the height of the solid body-like structure formed in the die cavity decreases. In the case of tableting, when the powder mass is confined in the die cavity, it cannot be expanded in the horizontal direction, leading to the generation of an additional force (i.e. radial force) acting on the direction perpendicular to the die-wall.

There are four different types of forces that act on the powder mass during compression. These are forces exerted by the upper punch acting in the downward direction, the force acting by the lower punch acting in the upward direction, die-wall frictional force acting in the upward direction and radial force acting perpendicular to the die-wall. The following relationship is present between die-wall frictional forces and radial forces.

$$F_D = \mu F_R \tag{6.1}$$

Where F_D is die-wall frictional forces, μ is the coefficient of die-wall friction and F_R is radial forces.

Frictional forces hinder the particle movement in the die-wall cavity due to the upper punch force which results in the non-uniform distribution of the forces in the die cavity. These forces depend upon the shear strength and contact area between the particle and die-wall. Shear strength is due to the adhesion strength present between the particle and the die-wall and the total contact area. Shear strength is the function of the bond between the particle and the die-wall. Lubrication of the die-wall results in weakening the bond by forming a layer between the particle and die-wall. Because the bond which forms between the lubricant and particle is weaker than the bond formed between the particle and die-wall. Furthermore, the total surface can be minimized by lowering the height-to-diameter ratio of the tablet.

6.5 Porosity

The porosity of the tablet can be defined as the ratio of the void volume to the bulk volume of the tablet. If void volume and bulk volume for the tablet are V_v and V_b, respectively, then

$$\text{Porosity } E = V_v / V_b \tag{6.2}$$

$$\text{Wc know that } V_v - V_b – V_t$$

$$\text{Therefore } \% \ E = [V_b – V_t]/V_b \times 100 = [1 - V_t/V_b] \times 100$$

The bulk volume of the tablet can be calculated by the geometric shape and true volume can be calculated from the true density and weight of the tablet. Porosity can define several processes including water uptake mechanism, swelling kinetics dissolution, and disintegration of the tablet in the GIT fluid which affects the in vivo behaviour of the tablet. Theoretically, an endpoint of the compression process is defined when the porosity becomes zero (i.e. $V_b = V_t$) and there is no void space available in the tablet. Practically, some porosity of the table is desirable for dissolution, disintegration, and swelling of the tablet. Therefore, the force-volume relationship in tableting during compression is critical in understanding the in vivo performance of the tablet.

Broadly, the maximum reduction in the bulk volume is due to the rearrangement of the particles under a high compression load. Plastic deformation or brittle fracture also contributes to the bulk volume reduction however to a lesser extent. The process of bulk volume reduction is not linear and uniform through the tablet structure. According to Heckel and his team, reduction of bulk volume during the compression follows first-order kinetics and pores can be considered as reactants in the processes. According to Heckel, at the beginning of the compression process, bulk volume reduction is very fast which is due to the filling of void space due to the movement of the particles in the die cavity. It becomes slow when voids between the particles significantly decrease and reach a threshold, beyond which no reduction takes place even an increase in the compression force. It explains the presence of some amount of porosity in the tablet. According to Heckel, porosity E can be defined as

$$\text{Log}\,(1/E) = K_yP + K_r \tag{6.3}$$

Where P is the applied load, K_y and K_r are constant representing the bulk volume reduction process due to deformation and rearrangement of the particle within the die cavity, respectively.

Heckel plots are used to identify the material based upon the deformation mechanism. Materials undergoing plastic deformation or brittle fracture during compression exhibit different plots (Niazi, 2009). For those materials which tend to undergo plastic deformation, final porosity depends upon the initial particle size and shape of the material. Materials of different particle size exhibit different porosity. On the other hand, in hard materials, which undergo brittle fracture, the final porosity does not depend upon the initial particle size. This is because the voids are filled by fines produced from fragments. Moreover, it is difficult to compress the hard material because the downward movement of the particles to fill the voids is extremely difficult (Fig. 6.2).

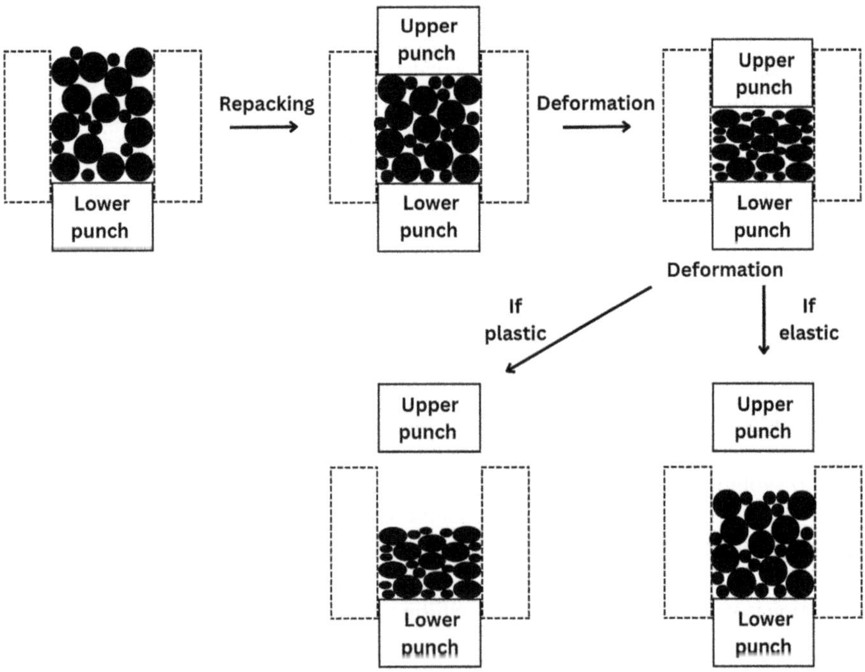

Fig. 6.2 Heckel plot representing different compaction regions

6.6 Ejection Forces

Ejection forces are those forces which are required to eject the tablet from the die cavity after the completion of a compression-decompression cycle. It mainly depends upon the radial and die-wall frictional forces generated during the compression. Higher the radial and die-wall frictional forces, higher the ejection forces. Both radial and die-wall frictional forces can be minimized by lubrication and reducing the height-to-diameter ratio of the tablet. It explains that lubrication eases the ejection of the tablet from the die-cavity after compression.

6.7 Decompression

Decompression is defined as the stage when compression load is removed. It is important to note that compressed powder mass in the die cavity still experiences the forces even after removal of the compression load although different in nature. During the decompression, forces arise from the relaxation of the elastic material and the ejection forces. If the forces generated from the elastic recovery of the material are more than the consolidation strength of the material, it may lead to structural failure for the tablet. It may be avoided by partial or complete replacement of elastic materials from the tablet formula. Therefore, the information obtained from the decompression cycle is also important to explain the quality of the tablet after the compression.

6.8 Compaction Profile

Compaction profile is a hysteresis curve obtained by plotting radial force versus axial force during the compression-decompression cycle. The purpose of the compaction profile of any tableting process is to find out the amount of residual radial pressure and rate of relaxation during the decompression cycle. It is used to measure the overall quality of the tablet after the compression. Radial forces are arisen due to the inability of the powder mass to expand horizontally in response to a vertical compression load due to the confinement of the powder mass in the die cavity. In the case of unconfined solid bodies, the ratio of change of vertical dimension to the horizontal dimension is used to characterize the material based upon the mechanical properties of the material. Although the tablet is not a perfect solid body, this ratio is used to explain the residual pressure after the compression. For the perfectly elastic bodies, materials regain their original state after the removal of the compression force, leading to the generation of internal forces. If the forces generated from the elastic recovery are higher than the consolidation strength of the tablet, structural failure may take place (Patel et al., 2006). In general, the compressional force used in

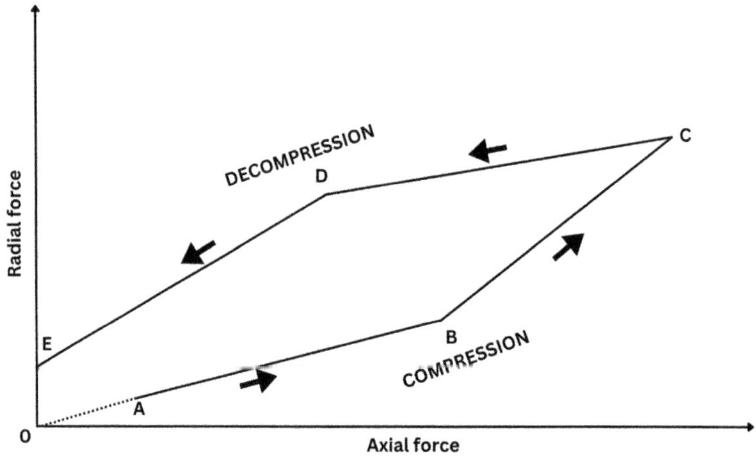

Fig. 6.3 Typical compaction profile observed during tableting

the tableting process is significantly higher than the elastic limit of the materials used in the formula. It results in plastic deformation or brittle fracture of the materials. In a typical compaction profile, the compression cycle is characterized by two distinct points, one, an elastic yield point, and second, a maximum axial compression force point. The slope of the curve is different in these two regions. In the decompression cycle, when maximum force is removed, an initial rapid change in the slope is observed, leading to obtaining a second yield point After the second yield point in the decompression cycle, the slope of the curve defines the residual radial pressure. A sharp change in the slope of the second stage of the compaction profile indicates fast recovery and low residual radial pressure which may lead to structural failure. The residual radial force also provides information about the ejection forces and lubricant requirement (Ghori & Conway, 2016). Overall, a compaction profile is used to define the structural quality of the tablet (Fig. 6.3).

References

Alderborn, G., & Aulton, M. E. (2002). *Pharmaceutics: The science of dosage form design*. Churchill Livingstone.

Carstensen, J. T. (2000). Advanced pharmaceutical solids. *Advanced Pharmaceutical Solids, 110*, 1–510. https://doi.org/10.1201/b16941

Ghori, M. U., & Conway, B. (2016). Powder compaction: Compression properties of cellulose ethers. *British Journal of Pharmacy, 1*(1). https://doi.org/10.5920/BJPHARM.2016.09

Jones, D. (2008). *Pharmaceutics – Dosage form and design*. Pharmaceutical Press.

Marshall, K. (1999). *Compression/Compaction*. FMC Corporation.

Kikuta, J., & Kitamori, N. (1983). Evaluation of the die wall friction during tablet ejection. *Powder Technology, 35*(2), 195–200. https://doi.org/10.1016/0032-5910(83)87009-0

Lachman, L., Liberman, H. A., & Schwartz, J. B. (1987). *Pharmaceutical dosage forms: Tablets* (Vol. 3). Marcel Dekker, Inc.

Lachman, L., Liberman, H. A., & Schwartz, J. B. (1989). *Pharmaceutical dosage forms: Tablets*. Marcel Dekker, Inc.

Niazi, S. (2009). *Handbook of pharmaceutical manufacturing formulations – Compressed solid products* (2nd ed.). CRC Press. https://doi.org/10.1201/b14437-93

Patel, S., Kaushal, A. M., & Bansal, A. K. (2006). Compression physics in the formulation development of tablets. *Critical Reviews in Therapeutic Drug Carrier Systems, 23*(1), 1–65. https://doi.org/10.1615/CRITREVTHERDRUGCARRIERSYST.V23.I1.10

Pazhayattil, A. B., et al. (2018). *Solid oral dose process validation*. AAPS Introductions in the Pharmaceutical Sciences.

Chapter 7
Pan Coating

Anil B. Jindal

Contents

Abstract Coating a tablet provides different features to the dosage form. Pan coating is the traditional method for film, enteric, or sugar coating of tablets. The coating process is dependent upon the advances in the techniques, equipment, and material used for coating. Despite advancements in the rules for scale-up of the pan-coating process, the laboratory scale process possesses several gaps leading to differences in the results observed on large scale. The chapter describes the thermodynamic factors, pan and spray-related factors that affect the pan-coating process, and determines the quality of the product. The final section of the chapter explains the scale-up of the pan-coating process.

Keywords Pan coating · Pan speed · Spray rate · Enteric coating

7.1 Introduction

The purpose of the coating of the tablet is to improve the elegance, protect the drug from degradation in the presence of moisture or oxygen if it is sensitive, mask the taste of the drug, and control the drug release (Lachman et al., 1987). Depending

A. B. Jindal (✉)
Department of Pharmacy, Birla Institute of Technology and Science, Pilani, Jhunjhunu, Rajasthan, India
e-mail: anil.jindal@pilani.bits-pilani.ac.in

A. B. Jindal (ed.), *Pharmaceutical Process Engineering and Scale-up Principles*, AAPS Introductions in the Pharmaceutical Sciences 13,
https://doi.org/10.1007/978-3-031-31380-6_7

upon the type of coating, the product is referred to as sugar-coated, film-coated, or enteric-coated. Sugarcoating is the oldest coating technique which is not commonly used nowadays. It is replaced by film coating, wherein a thin polymeric film is formed onto the surface of the tablet to perform several functions. On the other hand, enteric coating is also a type of film coating wherein a specific polymer is used to coat the tablet surface to prevent the disintegration of the tablet in the stomach and allows the dissolution of the coating in the intestine. The polymer used for the enteric coating is insoluble at acidic pH, thereby preventing the disintegration of the tablet in the stomach to protect against the degradation of the acid-labile drugs. The film coating process is accomplished by using the pan coater wherein a polymeric solution is sprayed onto the tablet moving in the pan coater, leading to the formation of a thin polymeric film onto the tablet surface after evaporation of the solvent. The chapter focuses on the variables which could affect the properties of the film. Scale-up of the pan-coating process has also been elaborated.

7.2 Effect of Different Factors on the Pan-Coating Process

Different factors which could affect the pan-coating process can be divided into three different categories, namely thermodynamic, pan-related, and spray-related factors (Pandey et al., 2014).

7.2.1 Thermodynamic Factors

Thermodynamic factors including temperature and humidity of the inlet and exhaust air and airflow rate exert a significant impact on the coating properties. Inlet air temperature and humidity play a key role in the rate of evaporation of the coating solvent from the tablet surface. Higher inlet air temperature may result in spray drying of the coating solution at the tip of the spray nozzle and blocking the nozzle. On the other hand, low inlet air temperature results in wet tablet surfaces and leads to twining of the tablets during coating. Moreover, the humidity of the inlet air defines the drying capacity of the air and hence the extent of drying of the tablet surface (Agrawal & Pandey, 2015). Therefore, the dew point of the air is useful for the determination of inlet air temperature during the coating process (Choi et al., 2021).

Exhaust air conditions are useful in determining the variations in the processing conditions within the coating pan. Batch-to-batch constant exhaust air temperature indicates similar processing conditions across the batches. Therefore, exhaust air temperature is monitored to understand the variation in the temperature and humidity inside the coating pan.

The airflow rate should also be kept constant across the batches to maintain the constant drying capacity to ensure the constant rate of solvent evaporation.

7.2.2 Pan-Related Factors

Different pan-related factors, including pan speed and pan load, affect the coating process. Pan speed defines the cascade movement of the tablet inside the pan and the duration for which a tablet remains in the spray zone during coating. The cascade movement of the tablet is very important to determine the quality of the film. Higher pan speed may result in improper cascading of the tablets inside the pan, leading to differential exposure of the tablet to the coating solvent and the formation of a polymeric film of different quality (Alderborn & Aulton, 2002). Another important parameter that can be affected by the pan speed is the dwell time of the tablet in the spray zone. Higher pan speed results in a lower dwell time, thereby higher process time. Therefore, an increase in pan speed during the coating is always accompanied by a proportional increase in the spray rate to reduce the processing time. An increase in the pan speed may also result in an increase in the friability of the core tablet during the coating due to the higher stress experienced by the tablet inside the pan.

Pan load or batch size is another pan-related factor which could exert a significant effect on the quality of the film. In general, a pan load of 60–80% of the total pan volume is considered optimum for the efficient coating process. Underloading the pan may result in spraying of coating solution onto baffles and pan walls, leading to an increase in wastage of the coating solution. Moreover, it also results in the formation of a polymeric film onto the pan wall surface which poses additional challenges in the cleaning of the pan. It may also result in the contamination of the next product. O the other hand, overloading of the pan may result in the improper cascade movement of the tablet inside the pan, leading to uneven exposure of the tablet to the coating solution and nonuniform film formation. At the laboratory scale, if the batch size is not sufficient to attain the optimum pan load, placebo tablets are added to increase the batch size for the coating (Fig. 7.1).

7.2.3 Spray-Related Factors

Spray-related factors, including spray rate, atomization air pressure, gun-to-bed distance, and spray pattern, are crucial to the quality of the film (Heinämäki et al., 1997). The spray rate is decided based on the drying capacity of the inlet air. An optimum spray rate is essential for uniform coating. Higher spray rate results in over-wetting of the tablet surface and twining and sticking of the tablets can be observed. It can be avoided by increasing the inlet air temperature. On the other hand, a low spray rate can increase the processing time due to the application of insufficient coating solutions. Other parameters which can be used for the optimization of spray rate are pan speed and pan load. The spray rate should be increased with an increase in the batch size and pan speed and vice versa.

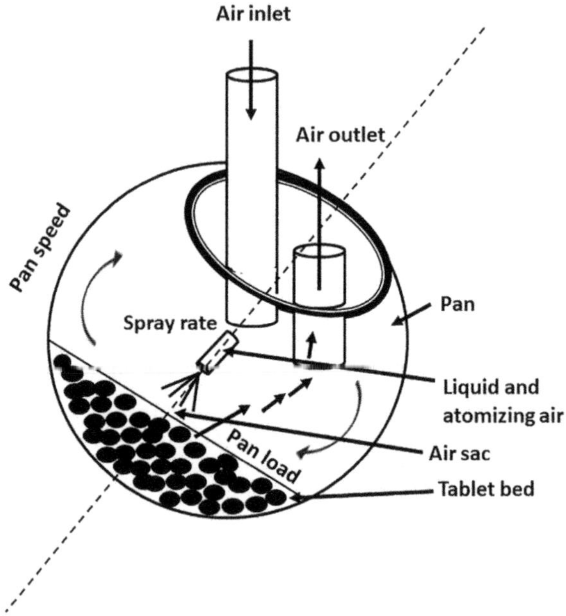

Fig. 7.1 Pan coater

Atomization air pressure defines the droplet size and the higher the atomization air pressure, the smaller the droplet size. In general, larger droplets may result in the dissolution of coating material on the tablet surface and the migration of the drug from the tablet to the film. While smaller droplet may be spray-dried before it reaches the tablet surface. It can be avoided by decreasing the inlet air temperature or reducing the gun-to-bed distance.

7.3 Scale-Up of the Pan-Coating Process

The following section describes the calculation of different parameters for the large-scale batch using optimized process parameters at the lab scale (Agrawal & Pandey, 2015).

7.3.1 Pan Speed

In general, at a large-scale coating pan speed is kept lower than the optimum speed to avoid additional mechanical stress onto the tablets during the initial stage of coating. Pan speed can be calculated by using either the Froude number approach or keeping the linear pan velocity constant across the scale. If the coating pans are geometrically similar across the scale, the Froude number approach can be used for the calculation of pan speed. In the case of geometric similarity, the ratio of centrifugal forces to inertial forces remains constant, represented by the Froude number. The following equation can be used for the calculation of the Froude number.

$$F = \mu^2 R / g.$$

The constant Froude number across the scales ensures the dynamic similarity in the process, indicating that a tablet in the coating pan experiences the same amount of centrifugal and inertial forces across the scales.

Another critical approach which is used to calculate the pan speed at a large scale is based on the calculation of the linear velocity of the tablets in the pan. At the laboratory scale, the linear velocity of the tablet is calculated considering the pan speed and dimensions of the pan. At the large scale, that pan speed is selected, which can maintain the linear velocity of the tablet equal to the laboratory scale (Pazhayattil et al., 2018). The following equation can be used for the calculation of pan speed at a large scale by this approach.

$$nD = \text{constant}$$

Constant linear velocity across the scale ensures the same dwell time for the tablet in the spray zone, hence similar coating properties.

7.3.2 Spray Rate

It is always desirable to use a maximum spray rate during the pan-coating process to minimize the production time. The spray rate during the pan-coating process depends upon the fraction of tablets (F) that remain in the spray zone during coating (Pandey et al., 2006). The F can be calculated by using the equation

$$F = n/N$$

n = the number of tablets remaining in the spray zone, and N = the total number of tablets in the coating pan.

Both the total number of tablets (N) and tablets present in the spray zone (n) at a specific time increase with an increase in the scale. However, an increase in N is significantly higher than the increase in n during the scale-up. In other words, tablets present in the spray zone decrease with an increase in the batch size. The spray rate is inversely proportional to the fraction of tablets present in the spray zone.

$$(\mathrm{SR})(\mathrm{F}) = \text{constant}$$

$$(\mathrm{SR})\ (n/N) = \text{constant}$$

SR is the spray rate.

n and N depend upon the pan diameter and pan load, respectively. Therefore

$$n/d = \text{constant}$$

$$d = \text{pan diameter}$$

$$N/\text{PL} = \text{constant}$$

$$\text{PL} = \text{pan load}$$

On the other hand, if there is no change in the environmental conditions, the spray rate is decided based on the airflow rate used during the process and can be calculated using equation

$$\text{SR}/V = \text{constant}$$

$$V = \text{airflow rate}$$

If there is a change in the environmental conditions during the technology transfer, thermodynamic principles are used for the spray rate calculations.

7.3.3 Pan Load

In general, pan loading is expressed in terms of the volume of the tablets rather than by weight of the tablets. If the pans used in different scales are geometrically similar, pan loading can be calculated using the equation

$$\text{Pan load}/\text{pan volume} = \text{constant}$$

Optimum pan loaded is desirable to avoid the losses of the coating solution and desired movement of the tablets in the pan. In production scale batches, the loading of the pan depends upon the batch size by weight and the capacity of the pan to hold the tablet weight per run. In this strategy, pan loading may be less than the optimum load to ensure the maximum utilization of the resources (Pandey et al., 2006).

7.3.4 Air Volume

Air volume plays a very important role in the pan-coating process. It decides the drying capacity of the air and hence the evaporation rate of the solvent. There are two approaches available which can be used to calculate the air volume during the scale-up process. Air volume in the large-scale batch can be decided based on the vendor recommendation for the coating pan. In general, it is desirable to use the maximum air volume recommended by the vendor, which allows for to use of the maximum spray rate and hence the minimum production time (Pandey et al., 2014). Another approach which can be used to determine the air volume is based on the spray rate. In this approach, first, the spray rate is decided by considering any of the following factors including batch size, pan speed, and a fraction of tablets present in the spray zone. In this approach, spray rate is calculated by keeping the spray rate to batch size ratio or spray rate to pan speed ratio constant.

7.3.5 Number of Spray Guns

In general, at a laboratory scale, the coating is accomplished by one gun because it is sufficient for efficient coverage of the bed. However, the number of guns is required to increase with an increase in the batch size to enable complete coverage of the tablet bed. Furthermore, an increase in the number of spray guns also results in a decrease in the processing time due to an increase in the total spray rate. How many spray guns should be used in the production scale batch will depend upon the type of gun available for the process. Ideally, for a successful scale-up, it is desirable to use the same spray gun which is used to produce a laboratory-scale batch (Agrawal & Pandey, 2015). However, the plant location is usually different from the R&D laboratory of the pharmaceutical companies, and the same type of gun sometime is not available at the plant. Nevertheless, irrespective of the type of gun used, it should not produce any kind of coating problems such as sticking or picking.

7.3.6 Gun-to-Bed Distance

The distance between the tip of the gun and the tablet bed surface decides the distance which must be travelled by the coating solution droplet before it reaches the tablet bed. An optimum distance is required for efficient coating. For example, if this distance is more, then it may result in spray drying of the solvent. On the other hand, if it is less, over-wetting of the tablet surface is possible. However, a higher distance is desirable because it produces a high spray area to cover the tablet bed. In laboratory-scale batches, the operator set this distance by using the ruler and visual observations by the naked eye. Moreover, in laboratory-scale batches, there is no scope available for the adjustment of this distance. While, in the case of product scale batches, due to the large coating pan, the operator needs to decide the gun position inside the pan for efficient coating (Agrawal & Pandey, 2015). There is no scientific strategy available which can be used to set the gun position. It mainly depends upon the practical experience of the operator.

References

Agrawal, A. M., & Pandey, P. (2015). Scale up of pan coating process using quality by design principles. *Journal of Pharmaceutical Sciences, 104*(11), 3589–3611. https://doi.org/10.1002/JPS.24582

Alderborn, G., & Aulton, M. E. (2002). *Pharmaceutics: The Science of Dosage form Design*. Pharmaceutics: The Science of Dosage form Design.

Choi, M., Porter, S. C., & Meisen, A. (2021). Interrelationships between coating uniformity and efficiency in pan coating processes. *AAPS PharmSciTech, 22*(8), 265. https://doi.org/10.1208/S12249-021-02155-Y

Heinämäki, J., et al. (1997). Optimization of aqueous-based film coating of tablets performed by a side-vented pan-coating system. *Pharmaceutical development and technology, 2*(4), 357–364. https://doi.org/10.3109/10837459709022634

Lachman, L., Liberman, H. A., & Kanig, J. L. (1987). *The theory and practice of industrial pharmacy*. Varghese Publishing House.

Pandey, P., et al. (2006). Scale-up of a pan-coating process. *AAPS PharmSciTech, 7*(4), E125. https://doi.org/10.1208/pt0704102

Pandey, P., Bindra, D. S., & Felton, L. A. (2014). Influence of process parameters on tablet bed microenvironmental factors during pan coating. *AAPS PharmSciTech, 15*(2), 296. https://doi.org/10.1208/S12249-013-0060-0

Pazhayattil, A. B., et al. (2018). *Solid oral dose process validation*. AAPS Introductions in the Pharmaceutical Sciences.

Chapter 8
Size Reduction

Anil B. Jindal

Contents

Abstract Size reduction refers to the process of the formation of small particles with a large surface area. The stress-strain curve of the material determines the energy needed for size reduction with a desirable particle size distribution. The study of equipment used for size reduction and their process parameters helps to reduce the energy supply. In this chapter, we have discussed the theories describing the energy required during milling, and the equipment used for milling including the ball mill, hammer mill, fluid energy mill, cutter mill, and oscillating granulator. Further, we have explained the factors affecting the size reduction process.

Keywords Size reduction · Milling · Fluid energy · Particle size distribution · Oscillating granulator

A. B. Jindal (✉)
Department of Pharmacy, Birla Institute of Technology and Science, Pilani, Jhunjhunu, Rajasthan, India
e-mail: anil.jindal@pilani.bits-pilani.ac.in

A. B. Jindal (ed.), *Pharmaceutical Process Engineering and Scale-up Principles*, AAPS Introductions in the Pharmaceutical Sciences 13,
https://doi.org/10.1007/978-3-031-31380-6_8

8.1 Introduction

Size reduction is an important unit operation in pharmaceutical dosage form design and can be defined as the process of reducing the size of pharmaceutical materials by mechanical methods. The particle size of the material plays a key role in the engineering of several pharmaceutical processes due to an increase in the total surface area of the particles (Lachman et al., 1987). For example, the surface area of a spherical particle of diameter 1 cm is 3.14 cm^2. If the same sphere is divided into 1000 L spheres of diameter 0.1 cm each, the total surface of the particle is now 31.4 cm^2. The total surface area of the particles is increased now almost ten-fold after the size reduction process. An increase in the surface area of the particles exerts a profound effect on the physicochemical properties of the drug and pharmaceutical unit operations. For instance, an increase in the surface area leads to a significant increase in the solubility of the drugs which may enhance the bioavailability of the drug after oral administration. An increase in the surface area due to the size reduction also results in an increase in the cohesive and adhesive forces onto the surface of the particles which create a significant impact on the powder flow, mixing, and compression. Fine materials are difficult to mix due to poor flow within the blender. Moreover, the presence of a high percentage of fines in the blend results in weight variation. Micronization of coarse active pharmaceutical ingredients improves the dissolution due to an increase in the drug solubility. The process of size reduction of particle size by mechanical means is referred to as milling (Alderborn & Aulton, 2002).

8.2 Theoretical Consideration of Milling Process

Depending upon the properties of the particles and strain applied, particles may undergo deformation or brittle fracture. Particle behaviour under mechanical stress can be explained based on the stress-strain relationship during the milling process. In the beginning, when particles experience some degree of mechanical force, stress and strain of the particles follow Hooke's law (i.e. stress is directly proportional to the strain to a certain degree of applied mechanical force) (Parrott, 1974). It is presented as the AB region in Fig. 8.1. In this region, after removal of the force, particles regain their original state, and such deformation behaviour of the particles is referred to as elastic. The slope of the linear portion of the stress-strain curve shown in Fig. 8.1 defines the degree of softness of the particles and can be determined by calculating Young's modulus. The force beyond which particles are not able to regain the original state is termed the yield point. After the yield point, particles do not regain their original state and such deformation behaviour is known as plastic in nature. In the plastic region of the stress-strain curve, the mechanical force at which particles fracture is represented by the fracture strength of the particles. The area under the curve of the stress-strain curve represents the energy required for the fracture and a measure of the strength of the particles.

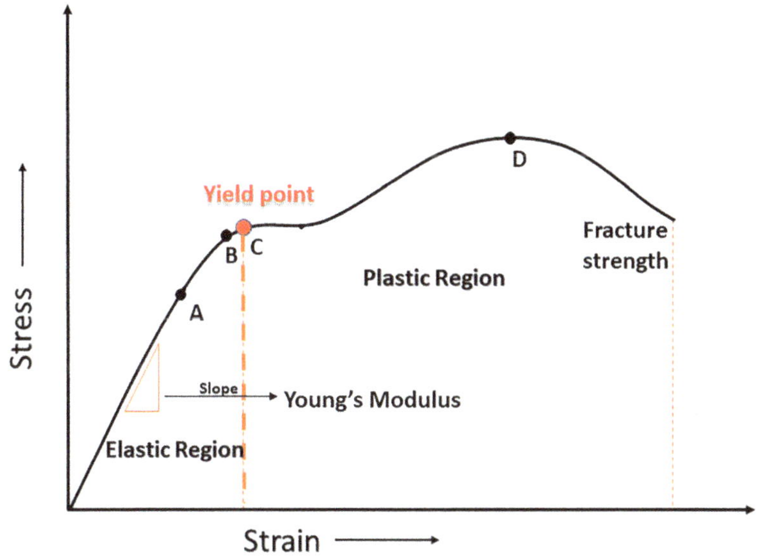

Fig. 8.1 Stress vs. strain curve

When a single particle is impacted by the hammer, it may divide into several particles of different sizes. If the degree of impact is increased in the process, although the number of small particles increases, size remains the same. It, therefore, appears that particle size distribution of the material after milling depends upon both the structure of the material and energy used in the milling process. Furthermore, the minimum size which can be obtained from a particular milling process depends upon the internal structure of the material (C.V.S Subramhanyam, 2001).

8.2.1 Energy Requirement in the Milling

1% of the total energy is used in the milling process and the remaining is lost in the other processes. The energy required for the size reduction of a material depends upon the increase in the surface area or a decrease in the particle size (Pazhayattil et al., 2018). The following theories have been suggested by different researchers for the calculation of energy requirements in a milling process. Each theory is based upon certain assumptions and validated in limited experimental conditions.

8.2.2 *Kick's Theory*

According to Kick and his team, the energy required for the size reduction of a material is directly proportional to the ratio of the diameter of the feed particles to the diameter of the product particle. If the diameter of feed and product particles are D_1 and D_2, respectively, the energy required in the process can be calculated using Eq. 8.1

$$E = C \ln(D_1/D_2) \tag{8.1}$$

where C is a constant which depends upon the crushing strength of the material and internal structure of the material.

Kick's theory considers the energy required for the deformation and brittle fracture of the material and ignores the presence of a distribution of flaws on the particle surface. According to Kick's theory, the energy required for the reduction of particle size of 1000 micron particles to 500-micron particles is the same as that required for reduction of particle size from 500 microns to 250 microns. It also contradicts the fact that the rushing strength of the particle increases with a decrease in the particle size, probably due to the presence of a few flaws on the small particle surface as compared to the large particles.

8.2.3 *Rittinger's Theory*

According to Rittinger and his group, energy required for the milling is directly proportional to the difference in the surfaces produced during the milling (Temmerman et al., 2013). If specific surface of feed and product material is S_1 and S_2 and energy required in the process is E, then

$$E = C' (S_1 - S_2)$$

where C' is a constant, which depends upon the relationship between particle diameter and surface area.

In terms of diameter,

$$E = C' (1/D_2 - 1/D_1)$$

Rittinger's theory assumes that energy required for the milling process is used in the generation of new surfaces and ignored the energy requirement for the deformation of the particles (Michaud, 2016). It is mainly applicable in the size reduction of brittle materials where little or no deformation is involved in the milling.

8.2.4 *Bond's Theory*

According to Bond and his team, the energy required forth the milling process is directly proportional to the square root of the diameter of the product particles. If the diameter of the product particles is D, then

$$W = 1/D^{0.5}$$

where W is the total work done in kilowatt-hours per short ton of the materials.

Bond's theory underestimates the role of feed material particle size in the milling efficiency, and it states that fine grinding requires more energy which is not dependent upon the particle size of the feed. For example, the energy required to produce a 1-micron particle from the 10 microns particles is the same as that to produce a 1-micron particle from the 5 microns particles. It also ignores the role of flaws in the milling process (Ganderton, 2014).

8.3 Ball Mill

Ball mill is used in the pharmaceutical industry to produce the fine particles of the drugs. It consists of a metal cylinder that can rotate on a horizontal axis and grinding medium (i.e. stainless steel or porcelain balls). Grinding of particles is accomplished by the impact and attrition of the balls in the rotating ball mill. Speed of the ball mill plays an important role in the size reduction process. At slow speed, balls in the mill exhibit cascade movement and particle size is reduced due to the attrition. On the other hand, when the speed of the mill is increased to an optimum speed, in addition to the attrition, balls also start moving up and fall on the material which provides an impact action for the size reduction. If the speed of the mill is increased beyond the optimum speed, it reaches the critical speed, and the balls start rotating with the walls of the mill due to the centrifugal force generated at a higher speed. Therefore, the critical speed of the mill can be defined as the speed at which the powder blender just starts centrifuging. At the critical speed, the gravitational force acting on the ball due to the weight of the ball is balanced by the centrifugal force (Fig. 8.2). If the mass of the ball is *m*, then, at critical speed

$$mg = mv^2/r$$

where *g* is gravitational acceleration, *r* is the radius of the mill and *v* is critical speed.

$$v = (gr)^{1/2}$$

If angular velocity of the mill is ω, then

$$v = \omega r$$

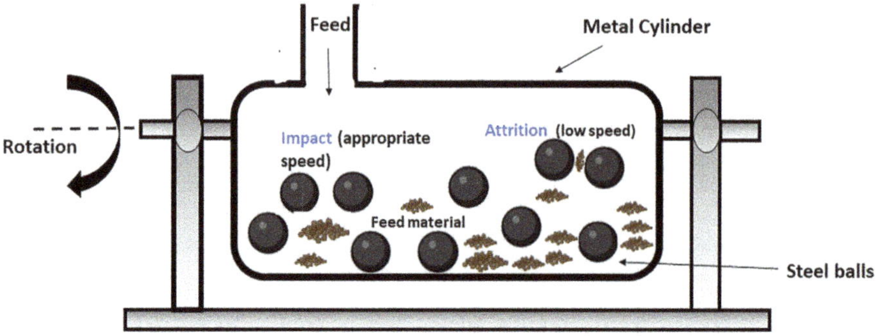

Fig. 8.2 Ball mill

In general, the optimum speed is 80% of the critical speed of the mill. The critical speed of the mill depends upon the size of the mill, and it can be calculated by the following equation

$$\omega = \left(g/r\right)^{1/2} \tag{8.2}$$

Calculation of the critical speed is useful in assessing the milling process (i.e. whether it is satisfactory or unsatisfactory) (Carter, 1996). For example, a ball mill of 1 m in diameter is rotating at 54 rpm. The critical angular velocity of the mill can be calculated as

$$\omega = (9.8/0.5)^{1/2}$$
$$= 4.42 \text{ radian per sec}$$

The critical speed of the mill is 43 rpm (calculated from critical angular velocity) (i.e. actual speed of the ball mill is more than the critical speed which indicates that the milling process is unsatisfactory).

Product particles are also affected by several other parameters including the size and weight of the balls, the volume occupied by the balls in the mill, and the amount of the material. In general, the particle size of the product depends upon the void volume present in the balls. Therefore, small balls produce small-sized particles as compared to larger balls due to the less valid volume available for particle movement during the grinding. Moreover, small balls exert more impacts per unit weight of the material due to the increase in the number of balls, which also contribute to the size reduction of the material.

In general, balls should not occupy more than half of the mill volume. If the volume occupied by the balls is significantly high, movement in the balls may be hindered, leading to insufficient attrition and impact effect for the size reduction and milling efficiency decreases.

Furthermore, the total weight of balls produces fine particles, and the total weight can be increased either by increasing the number of balls or by increasing the density of the ball. An increase in the number of balls results in an increase in the volume of the mill occupied by the balls which should not be more than half of the total volume of the mill. Therefore, balls made up of dense material are used to produce fines. For example, steel balls are denser than porcelain balls and are used to produce fine material.

The ball mill is very versatile in operation and can be used for both wet and dry milling. However, it produces a lot of noise due to the movement of collision of the metal balls with the walls of the mill during grinding. Another limitation of the ball mill is that it may contaminate the product due to the abrasion of the metal from the surface of the balls during the attrition which could be a very serious concern in the manufacturing of pharmaceutical products. However, balls made up of stainless steel can be used for the manufacturing of parenteral products due to less contamination.

8.4 Hammer Mill

Hammer mill is widely used in the pharmaceutical industry for the milling of various drugs. A typical hammer mill consists of a few hammers mounted on a rotor, which can rotate at a specific speed, leading to the swinging of the hammer in the mill and enclosed in a metal housing. An output perforated screen is present at the bottom of the mill which plays a significant role in the determination of the particle size of the product. When feed material enters the mill, feed particles start moving in the mill under the influence of gravitational and centrifugal forces. Size reduction is accomplished by the impact of rotating hammers on the moving particles. The degree of impact is significantly influenced by the speed and direction of moving particles. Particles generally move in the direction of the hammer at a speed that is higher than the hammers. The higher speed of the moving particles is due to the presence of additional drag force exerted by the hammers on the particles. To overcome this problem, deflectors are introduced in the mill, which alters the path of the particles and increases the impact of the hammer. Speed of hammers also plays an important role in the size reduction process. At significantly low speed, blending rather than milling effect is obtained from the rotating hammers. A maximum milling efficiency is obtained at the critical speed. Significantly higher speed is also not suitable as particles come out from the mill through the screen immediately after it reaches a size lower than the aperture size of the screen. Output screen size and thickness are also critical in determining the particle size of the product. The major role of the output screen is to retain the particles in the mill until the desired size is not achieved. In general, product particle is always less than the output screen size (Fig. 8.3). For example, if a 30 mesh output screen is used in the mill, the particle size of the product material is always less than the 30 mesh size. The particle size of the discharged material mainly depends upon the velocity of the particle and the angle at which it exits from the mill. As particles exit from the mill tangentially, the size of

Fig. 8.3 Hammer mill

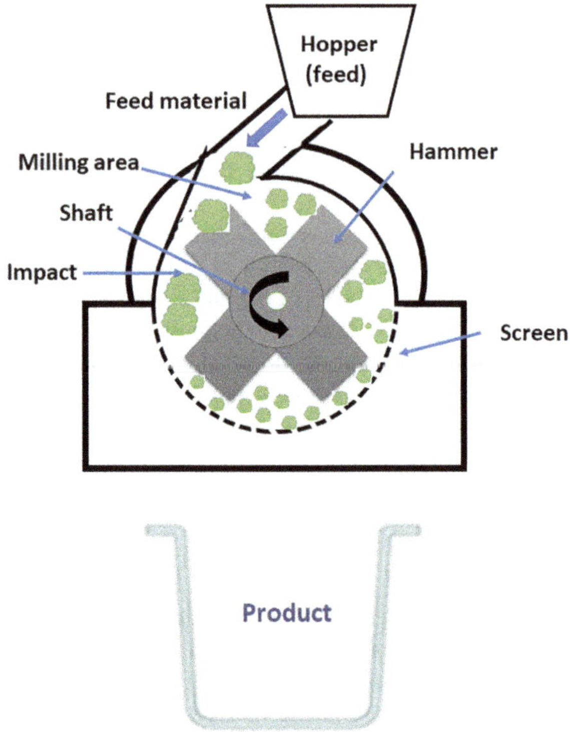

the particles is always less than the output screen size. Furthermore, as the particle velocity inside the mill increases, the exit angle of the particles decreases, leading to a decrease in the discharge material particle size. The thickness of the output screen also affects the discharge material particle size and particle size decreases as the output screen thickness increases. An important advantage of the hammer mill is that a product with a narrow particle size distribution can be obtained. This is because the particle exits the mill as soon as it reaches a size less than the aperture size of the output screen. Generally, a hammer mill can produce particles of 20 to 40-micron sizes. Small particles have less inertial forces as compared to large particles, leading to a lower impact generated by the size reduction (Hickey & Ganderton 2016).

8.5 Fluid Energy Mill

A fluid energy mill can produce extremely fine particles and is used for the micronization of active pharmaceutical ingredients in the pharmaceutical industry. Particles enter the mill from the bottom and air is supplied through a nozzle into the mill. After entering the mill, particles drag by the air and adapt the elliptical path. Size reduction is accomplished by the interparticle collision or collision of the

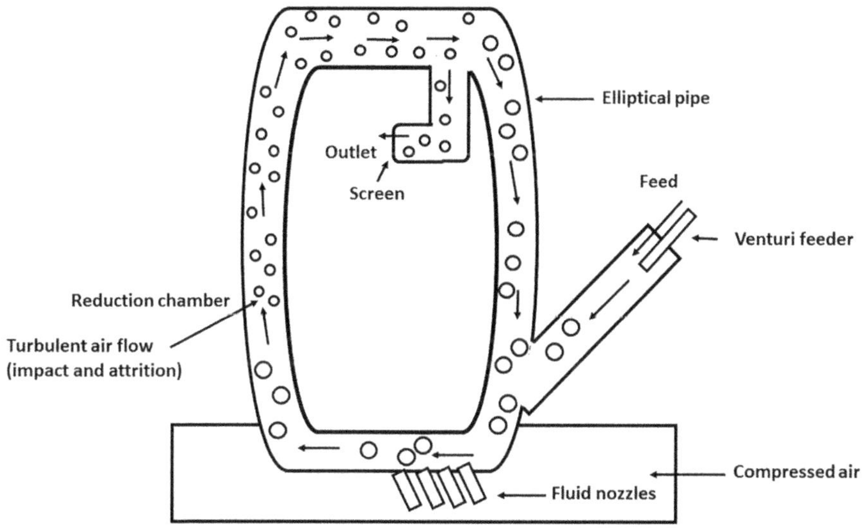

Fig. 8.4 Fluid energy mill

particles with the walls of the mill. The mechanism of particle size reduction is impact and attrition. Unlike the ball mill, there is no grinding medium is used in the fluid energy mill for the size reduction of the materials. Due to the expansion of the gas in the chamber, temperature in the mill decreases which enables the use of the fluid energy mill for the milling of thermolabile substances (Lachman et al., 1987) (Fig. 8.4).

Feed rate is very critical for the size reduction of the materials. At a high feed rate, the number of particles in the mill increases which results in a decrease in the mean free path length between the particles (i.e. distance travelled by the particles before the collision). A decrease in the mean free path leads to a decrease in the kinetic energy and impact during a collision which produces the large particles. On the other hand, at a low feed rate, when the mean free path is more, small particles are produced. Based upon the particle size, particles are entrained from the mill by the cyclone separator. When particles are in the mill, it experiences two forces, a drag force exerted by the air which moves the particle in the forward direction and a centrifugal force for moving the particles towards the periphery of the mill. Therefore, large particles remain in the mill until a sufficient small particle size is obtained which can be entrained from the mill (Pharmapproach, 2022).

A fluid energy mill offers several advantages over a ball mill. (i.e. there is no contamination of the product with metal particles and it is also suitable for milling of thermolabile substances). It does not produce any noise during the milling process and cleaning and maintenance are also easy.

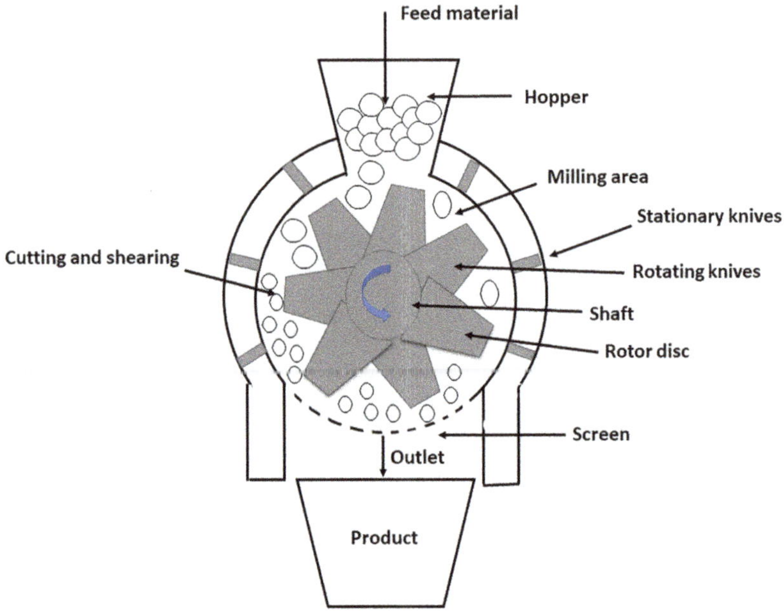

Fig. 8.5 Cutter mill

8.6 Cutter Mill

A cutter mill is used for the size reduction of soft materials (i.e. plant parts) and produces coarse particles. It consists of a sharp knife that is mounted on the shaft and the entire assembly is enclosed in the metal shell. Size reduction is accomplished by the cutting mechanism of the knife. Similar to the hammer mill, an output screen is present at the bottom of the mill which allows the discharge of the materials with a particle size smaller than the aperture of the screen. Different factors such as speed of the knife and feed rate which are listed in the hammer mill also affect the milling process in the case of the cutter mill. It is mainly used to grind the plant materials before the extraction to decrease the particle size (Carter, 1996) (Fig. 8.5).

8.7 Oscillating Granulator

It is widely used in the pharmaceutical industry for the wet or dry milling of granules in tableting before lubrication and compression. It consists of an oscillator with a mesh. The mesh is available in different sizes depending upon the granule size required. The purpose of the oscillating granulator is to obtain the granules of the narrow size distribution for uniform drying of wet granules. On the other hand, in the case of milling dry granules, the purpose is to obtain the granules with narrow

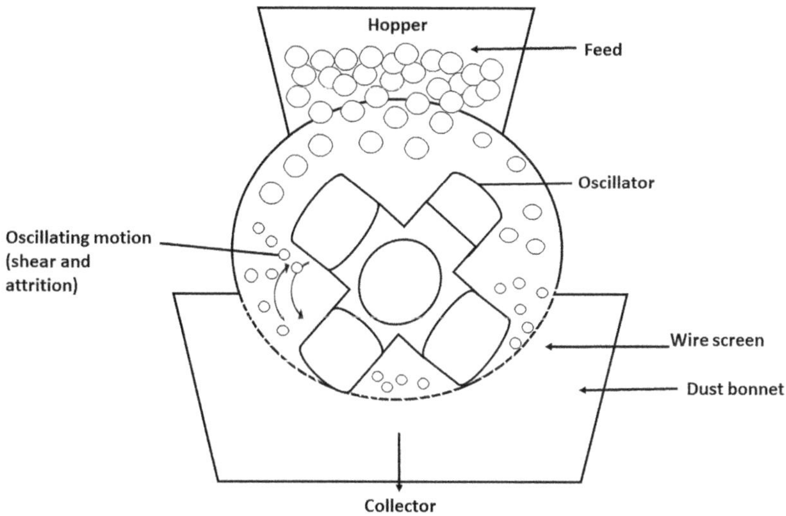

Fig. 8.6 Oscillating granulator

particle size distribution for uniform lubrication. The shear is the main mechanism of the particle size reduction in the oscillating granulator (Fig. 8.6). Although it is a very robust mill, metal contamination of the product and high wear and tear is the major limitations of the mill. Another disadvantage of the oscillating granulator is that it produces a high amount of fines from the grinding of brittle materials which may impart additional challenges during compression.

8.8 Factors Affecting the Size Reduction Process

Several materials-related factors including hardness, toughness, stickiness, and abrasiveness produce a significant effect on the size reduction process. It is more difficult to grind soft and tough material than hard and brittle. Most of the fibrous materials are soft in nature and difficult to grind. It can be converted into a brittle substance by treating it with liquid nitrogen. However, it is a costly process and practically not feasible (Hickey & Ganderton 2016).

The abrasiveness of the material to be milled is also a serious concern. It may lead to contamination of the milled material with the metal surfaces of the equipment.

Few substances are sticky in nature and tend to adhere to the surface of the mills during milling. It may be converted into free-flowing powder by adsorbing onto an inert material which facilitates the milling process.

Heat is generated during the grinding of materials from most of the mills. For example, a ball mill is not suitable for the milling of thermolabile substances. The temperature may rise in a hammer mill at a high feed rate. An increase in temperature may result in the softening of the few low-melting substances during the milling.

The presence of moisture increases the softness of the materials and drying of such materials before milling may enhance the milling efficiency. Moreover, depending upon the amount of moisture present in the material to be milled, a dry or wet milling process should be used.

8.9 Selection of the Mill

The selection of the mill for the size reduction of pharmaceutical substances depends upon several factors. For example, the particle size of the product, the nature of the material to be milled, and the heat sensitivity of the substance.

Ball mill and fluid energy mill can produce fine particle size and is used for the micronization of poorly soluble active pharmaceutical ingredients to formulate immediate-release solid dosage form (Carter, 1996). However, a fluid energy mill is preferred over a ball mill for the grinding of thermolabile substances. Cutter mill is generally used for the milling of soft plant material before extraction. A closed mill is generally used for the milling of therapeutically potent substances to avoid excessive exposure of the powder dust generated during the milling to the operator.

References

Alderborn, G., & Aulton, M. E. (2002). *Pharmaceutics: The science of dosage form design*. Churchill Livingstone.

Hickey, A. J., & Ganderton, D. (Eds.). (2016). *Pharmaceutical process engineering*. CRC Press.

Carter, S. J. (1996). *Cooper and Gunn tutorial-pharmacy*. CBS Publishers.

Ganderton, D. (2014). *Unit Processes in Pharmacy: Pharmaceutical Monographs (Vol. 7)*. Elsevier.

Lachman, L., Liberman, H. A., & J. L. K. (1987). *The theory and practice of industrial pharmacy*. Varghese.

Michaud, L. D. (2016). *Rock crushing theory and formula using Kick & Rittinger's law*. Crushing & Screening, Hardness, Laboratory Procedures. https://www.911metallurgist.com/blog/rock-crushing-theory-formula-using-kick-rittinger-law. Accessed: 4 Oct 2022

Parrott, E. L. (1974). Milling of pharmaceutical solids. *Journal of Pharmaceutical Sciences, 63*(6), 813–829. https://doi.org/10.1002/JPS.2600630603

Pazhayattil, A. B., et al. (2018). *Solid Oral dose process validation*. AAPS Introductions in the Pharmaceutical Sciences.

Pharmapproach. (2022). *Fluidized energy mill: Operating principles*. Available at: https://www.pharmapproach.com/fluidized-energy-mill/. Accessed: 9 Jan 2023

Subramhanyam, C. V. S. (2001). *Pharmaceutical engineering – Principles and practices*. Vallabh Publisher.

Temmerman, M., Jensen, P. D., & Hébert, J. (2013). Von Rittinger theory adapted to wood chip and pellet milling, in a laboratory scale hammermill. *Biomass and Bioenergy, 56*, 70–81. https://doi.org/10.1016/J.BIOMBIOE.2013.04.020

Part III
Liquid Dosage Forms: Pharmaceutical Process Engineering and Scale Up

Chapter 9
Mixing and Filtration

Dhwani Rana, Rikin Patel, Amit Sharma, Sagar Salave, and Derajram Benival

Contents

Abstract Mixing and filtration are crucial in the pharmaceutical industry. Poor mixing can result in product failure, while sterile filtration is the only method to ensure sterility for certain drug products. This chapter will cover the types of mixtures, solid-liquid mixing in pharmaceutical suspensions, and the equipment used for their development. Additionally, it will discuss the importance of sterile filtration for products that cannot undergo terminal sterilization. The goal is to provide an overview of these two essential unit operations.

D. Rana · A. Sharma · S. Salave · D. Benival (✉)
National Institute of Pharmaceutical Education and Research – Ahmedabad (NIPER–A) An Institute of National Importance, Government of India, Department of Pharmaceuticals, Ministry of Chemicals and Fertilizers, Gandhinagar, Gujarat, India
e-mail: derajram@niperahm.res.in

R. Patel
Intas Pharmaceuticals Ltd., Ahmedabad, Gujarat, India

A. B. Jindal (ed.), *Pharmaceutical Process Engineering and Scale-up Principles*, AAPS Introductions in the Pharmaceutical Sciences 13,
https://doi.org/10.1007/978-3-031-31380-6_9

Keywords Types of mixture · Solid-liquid mixing · Suspensions · Sterile filtration · Filter validation

Abbreviations

EMA	European Medicines Agency
MWCO	Molecular weight cut-off
PTFE	Poly (tetrafluorethylene)
PVDF	Poly (vinylidene fluoride)
US FDA	United States Food and Drug Administration

9.1 Introduction

This chapter delves into the critical role of mixing and sterile filtration in the pharmaceutical industry. Mixing is a fundamental unit operation used in the production of various dosage forms, as it ensures that different components are evenly combined and distributed. On the other hand, sterile filtration is essential in the manufacturing of sterile drug products that cannot be sterilized through terminal methods. The chapter focuses on the principles and techniques of these unit operations, including the various equipment and methods used to achieve optimal mixing and sterilization. The goal of the chapter is to provide a comprehensive understanding of the importance of mixing and sterile filtration in the pharmaceutical industry.

9.2 Types of Mixtures

Mixing enables the formation of three types of mixture: positive mixtures, negative mixtures, and neutral mixtures. If the participating components mix spontaneously and irreversibly by the process of diffusion and result in a homogeneous mixing, then this type of mixture is called a positive mixture. Generally, the mixing of gases or miscible liquids results in the formation of positive mixtures. It is important to note that the formation of a positive mixture does not require the input of energy, and it is only a matter of time which decides the degree of mixing. It means that if the time available for mixing is unlimited, it will result in a high degree of homogeneous mixing, and the input of energy can reduce the time required to achieve the same level of mixing. Negative mixtures are those in which the participating components tend to get separated after uniform mixing. Suspensions and emulsions are some examples of negative mixtures. Importantly, unlike a positive mixture, the formation of a negative mixture requires the input of energy. In addition, to maintain the uniformity of the mixture, a continuous input of energy is required, as in the case of suspension, where the moment stirring stops, sedimentation of particles will result, depending on the particle size of the suspended particles. On the other hand, neutral mixtures are those in which, similar to negative mixtures, an input of energy is needed to achieve homogeneous mixing; however, unlike negative mixtures, once mixed, they do not tend to undergo spontaneous segregation (Twitchell, 2012). From this, it is evident that mixtures such as powder, cream, and paste are some examples of neutral mixtures in which energy is required for their formation, and once they are formed, due to their static behaviour, an input of energy is not required to maintain the uniform mixing. Powder mixing is not the scope of this chapter as it is already discussed in the previous chapter, and solid-liquid mixing and liquid-liquid mixing, specifically oil-aqueous mixing, is discussed in this chapter.

9.3 Solid-Liquid Mixing

The solid-liquid mixing process for suspensions remains a very important unit operation for the pharmaceutical industry. Suspensions, creams, and pastes are just a few examples of dosage forms that involve solid-liquid mixing. Additionally, there are some dosage forms where the final product is solid. Still, during the manufacturing process, one of the intermediate unit operations involves solid-liquid mixing for example, loading an aqueous dispersion of the drug onto non-pareil seeds in a Wurster coater and filling these dried drug-loaded pellets either in capsule form or compressed into a tablet dosage form. This section of the chapter will focus on the solid-liquid mixing process for suspensions and will provide an in-depth understanding of the principles and techniques involved in this process.

9.3.1 Pharmaceutical Suspension

9.3.1.1 Wetting of Solid

When the solid particles (dispersed phase) are added into the liquid media (dispersion media), the uniform mixing of the solid particles into the liquid media depends on many factors. This includes the physicochemical properties of solid particles as well as liquid medium, additives included in the dispersion media, as well as hydrodynamic conditions in the processing unit, such as a manufacturing tank with a stirrer. Generally, achieving uniform mixing would be easy for a powder which has density greater than the liquid, the dispersion media has the ability to wet the solid particles, and the powder does not contain strongly bound particle agglomerates (Etchells AW, 2001; Özcan-Taşkin, 2015). However, this is not the case for drug substances, as most of the drugs are hydrophobic in nature. Generally, the dispersion media is water, so particle wetting remains a concern. To achieve uniform mixing, it is essential that when such hydrophobic powdered drug substance is added, it must first be separated into individual particles. For this, individual particles must be wetted. To facilitate ease of wetting, surfactants are included in the aqueous media. Surfactants promote the wetting of hydrophobic particles by reducing the contact angle (the angle the liquid makes with the solid surface) (Kolhe et al., 2013). Values of the contact angle between 0° and 90° suggest partial wetting, greater than 90° as nonwetting conditions. In comparison, a contact angle of 0° suggests complete wetting, and an angle close to 180° is unwettable by the liquid in question.

9.3.1.2 Mixing Uniformity

Wetting the powdered drug substance alone does not guarantee uniform mixing in the liquid dispersion media. Uniform mixing is defined in terms of the degree of dispersion. When the powdered drug substance, after wetting, releases the individual particles, it results in a very high degree of uniformity. It is to be noted that there may be a few aggregates that are made up of two or more particles; however, such aggregates would be relatively few in this case. Achieving this uniform mixing relies on the usage of different kinds of equipment. Mixing equipment is known by different names, and the most used name include mixer, agitator, stirrer, and blender. However, among these various names, for suspension, agitator would be more appropriate as it not only helps achieve the initial uniformity but also maintain the uniformity by providing the required agitation. Although blenders are also used to achieve uniformity, they are used whenever we want to mix solid powder components.

There are different types of equipments available for mixing purposes, and each uses different methods to achieve uniform mixing. Therefore, it is paramount that the selection of a particular mixer should be based on the desired purpose rather than trial-based approach. In stirred vessels, flow patterns are extremely important to achieve and maintain a high degree of mixing. To accomplish this, motion must be provided to entire contents of the tank. A swirling flow pattern and deep vortex on the surface are not signs of an effective mixing (Dickey, 2015). This circular motion moves the same dispersion within the tank and around the rotating impeller without any significant vertical or radial mixing. Depending on the type of mixing equipment, it can produce minimal shear, making it suitable for mixing two liquids. At the same time, high-shear homogenization would be required to break the agglomerates of tightly bound drug particles. Generally, equipment producing moderate shear is used at the laboratory scale, while high-shear devices, such as homogenizers, are used at the production scale. For suspension, besides the selection of equipment to achieve uniform dispersion, it is also equally important to consider three parameters, which include the selection of the most appropriate vehicle, avoiding air entrapment during the addition of various components needed in the preparation of suspension, and the use of a pre-dispersion technique to improve the initial degree of dispersion. Since water remains the vehicle of choice for most pharmaceutical suspensions as a dispersion media and drug substances are highly hydrophobic, the technique of pre-dispersion helps in overcoming the initial dispersibility issue whenever the surfactant is not desirable. In this technique, the drug substance is first dispersed in a small amount of liquid, such as glycerine, and this pre-dispersion is then added to the final aqueous dispersion media. The following section discusses various equipment used for the manufacturing of solid-liquid-based suspensions with the ultimate objective of achieving and maintaining high uniformity of dispersion.

9.3.1.3 Mechanically Stirred Vessel

The simplest equipment for manufacturing suspension is a mechanically stirred vessel or simply an agitated tank. In the laboratory, the most common mixer is a magnetic stirrer with a stir bar, which is generally used in a beaker/flask. Occasionally, this magnetic stirrer is part of a hot plate. Another typical laboratory mixer is a rotating shaft with a simple propeller/impeller (Dickey, 2015). The fundamental mechanism involved in this type of equipment is the physical movement of material between different parts of the entire mass using rotating impeller blades (Hemrajani & Tatterson, 2003). For laboratory-scale mixers, the mixing time is generally very short, and uniform mixing is achieved quite rapidly. However, this may not be the case for large production scale mechanically stirred tanks, and hence, to achieve uniform mixing, sometimes baffles are also positioned into the inner wall of the tank. The positions of these baffles are set in such a way that it transforms tangential flows generated by the rotating impeller into vertical flows and provide top-to-bottom mixing without swirl. In addition, baffles also minimize air entrainment. In large production-scale tanks, the impeller can be introduced into the tank from the top, bottom, side-entered, or angular top entering (10–15° angle from the vertical). In addition to the main impeller, occasionally, a small secondary impeller is also installed at the bottom of the tank. This secondary small impeller, which is also called a *tickler* or *kicker*, provides agitation when the liquid level falls below the main impeller level. Although many impeller variants are available in the market, broadly, almost all of them can be grouped as either turbines for the processing of low to medium-viscosity fluids or close-clearance impellers for the processing of high-viscosity fluids. Further, based on the flow patterns and level of shear generation, turbine impellers can be classified as axial flow, radial flow, hydrofoil, and high-shear impellers. Since suspensions tend to undergo sedimentation, it is important to provide liquid velocities at the floor of the tank for an effective sweeping action (Hemrajani & Tatterson, 2003). Axial flow impellers are recommended for such applications.

9.3.1.4 Rotor-Stator Mixing Devices

The unique feature of a rotor–stator mixer is a high-speed rotor near a stator. As the local energy dissipation and shear rates generated in a rotor-stator mixer are very high in comparison to those in a mechanically stirred vessel, and hence such mixers

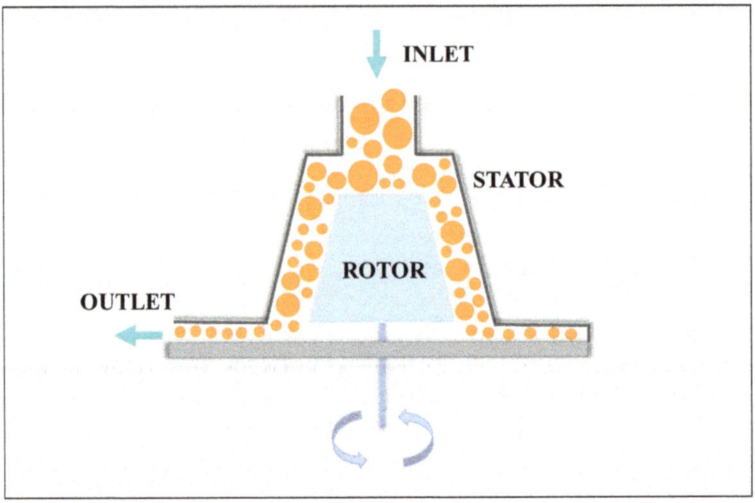

Fig. 9.1 Operating principle of colloid mill

are also called high-shear mixers (Atiemo-Obeng & Calabrese, 2003). The processing capacity of the rotor-stator mixer ranges from a few mL for small lab models to >1000 L/min of flow rates for large production units. Because such mixers are not only efficient in achieving uniform mixing but also capable of reducing particle size and hence, they are often also called a generator. Such mixers can be combined with the mechanically stirred vessel to enable a high degree of uniformity by liberating the particles from the strongly agglomerated tiny lumps. There are mainly three types of rotor-stator mixing devices available commercially, including Colloid Mills and Toothed Devices, Radial Discharge Impellers (Silverson Machines or Ross type), and Axial Discharge Impeller (Chemineer Greerco rotor–stator mixer). The schematic illustration of the working mechanism in a colloid mill is shown in Fig. 9.1.

Depending on the viscosity of the final formulation, many suspensions require either continuous or periodic agitation during the filling unit operation. If there is a delivery line between the bulk storage tank and the filling equipment, some segregation may occur, specifically if the suspension is not vicious. Therefore, good manufacturing practice demands testing vials/bottles at the beginning, in the middle, and at the end of the filling activity to ensure that no segregation has occurred. Notably, such samples from the initial, middle, and end should not be composited or pooled. Unlike solid mixing, where overmixing may result in segregation, intense agitation may lead to air entrapment, particularly for viscous suspensions. This air entrapment can impact not only the degradation rate for oxygen-sensitive drug products but also affect content uniformity from vial to vial. Hence, the agitation intensity should be periodically adjusted depending on the volume left behind in the bulk storage tank.

9.4 Liquid-Liquid Mixing

Mixing miscible liquids can be easily performed with simple equipment such as a mechanically stirred vessel, as discussed above, to mix solid-liquid suspension. However, mixing two immiscible liquids, such as oil and water, poses a challenge. Mixing two immiscible liquids results in the formation of a liquid-liquid dispersed system and such systems can be categorized into two types: dispersion and emulsion. In the case of dispersion, one phase forms a drop in the size range of tens to hundreds of microns when mixed in a stirred vessel, and the continuous input of mechanical energy stabilizes it. On the other hand, emulsions are described as liquid-liquid dispersed systems containing an emulsifier(s) with globule size/drop size of a few microns or lesser than this and stabilized by surface energy. Importantly, contrary to dispersions, emulsions remain stable for a few days to many months. However, dispersions are used in the pharmaceutical industries for extraction purposes as they tend to separate quickly into two layers once the energy input ceases. This chapter discusses pharmaceutical emulsions because they remain the final dosage form, and hence, mixing uniformity is very important. Different emulsions, such as water-in-oil(w/o), oil-in-water(o/w), and multiple emulsions, such as w/o/w or o/w/o, can be prepared. Considering this wide variety of formulations in composition and the strict regulations required for manufacturing, emulsification processes are usually based on experience (Pacek, 2015).

9.4.1 *Equipment for Manufacturing of Emulsions*

For the preparation of an emulsion, first, the oil phase and the aqueous phase are prepared separately, and then both these phases are mixed. This mixing is almost, in every case, first done in conventional mechanically stirred vessels, as discussed above, to prepare a suspension. The resulting emulsion is subjected to further processing equipment depending on the desired globule/drop size.

9.4.1.1 Mechanically Stirred Vessel

A mechanically stirred vessel has a rotating impeller in a stationary cylindrical tank. Impellers are characterized by power numbers, *Po. It is* a coefficient which relates the energy dissipation rate to the density of liquid and diameter as well as the RPM of the impeller. Traditionally, for efficient processing of liquid/liquid dispersions, radial flow Rushton turbines (which are also called 'high shear impellers') were considered. Low *Po* impellers like axial flow hydrofoils (Chemineer HE3) and special 'ultra-high shear' impellers such as Chemineer CS2 and CS4 are developed for producing fine emulsion (Pacek, 2015).

9.4.1.2 Rotor-Stator Mixing Devices

Rotor-stator mixing devices, which are also known as high-shear mixing devices, are widely used in the manufacturing of stable pharmaceutical emulsions. Various types of rotor-stator mixing devices are available commercially. However, all mixing devices have one common feature; a high-speed rotor which is closely surrounded by a stator. The gap between the rotor and stator can be adjusted depending on the product requirement and it is usually in the range of 100–3000 μm (Karbstein & Schubert, 1995). This gap results in a shear rate from 20,000 to 1,00,000 s^{-1} (Atiemo-Obeng & Calabrese, 2003). Further, the kinetic energy provided by the rotating rotor dissipates in a very small volume around the stator. As a result, the local energy dissipation rate is very high, in the range of a10^4 to 10^6 W/kg (Utomo et al., 2009). Therefore, the combination of a very high shear rate along with a very high energy dissipation rate enables the formation of emulsions with globules/drops in the size range of hundreds of nanometres to a few microns (Pacek, 2015).

9.4.1.3 High-Pressure Homogenizers

Homogenization is the process of narrowing or "homogenising" the particle-size distribution of an emulsion or a suspension, hence lowering the polydispersity of the sample. Particle breaking in the homogenizer is accomplished by a combination of high shear, turbulence, impact, and cavitation. The type of homogenizer employed, as well as the physical attributes of the material, decide which of these mechanisms is more essential for particle-size reduction (Morales et al., 2016).

9.5 Filtration

Filtration, a separation unit operation, is widely used in the pharmaceutical and biopharmaceutical industries to remove particulate contaminations, including microorganisms, from liquids, air, and gases. Even for a drug product to be terminally sterilized by the thermal method, the removal of particulate matter has dual implications; (a) it imparts a high degree of clarity to the solution, which is desirable for a parenteral solution and conveys the impression of high quality; (b) removal of microbial contamination helps in eliminating elevated levels of endotoxins, the debris of gram-negative organisms, and is responsible for producing fever.

9.5.1 *Classification of Filters*

Various types of filters differing in pore size and porosity band are available for different applications, and they are briefly discussed below;

9.5.1.1 Particle/Clarifying Filters

Particle filters have porosity ranging from 10 to 200 μm and are used to remove dirt, most particles, and some bacteria. Depth filters made up of highly adsorptive cellulosic or kieselguhr in the lenticular filter, design formats are widely used as they offer high dirt load capacity. These highly adsorptive filters find applications in the processing of plasma and serum samples to separate colloidal substances and lipids. In addition, such filters are now widely used in biopharmaceutical industries in cell harvesting to separate cell debris and other contaminants after the completion of the fermentation process. It is to be noted that asbestos was widely used for decades as a depth filter. However, usage is now banned due to its carcinogenic properties. A prefilter is a type of depth filter usually constructed of nonwoven or melt-blown fibre materials such as polypropylene, polyamide, cellulosic, sintered stainless steel, and glass fibre (Levy & Jornitz, 2006). These treatments eliminated the fibre shedding concern observed with the woven material-based depth filters. Prefilters are also placed before sterilizing the grade membrane filter to protect the primary filter from early clogging. Unlike sterilizing grade membrane filters, which have an absolute rating, prefilters have a nominal rating. The concept of rating in terms of particle retention is discussed later.

9.5.1.2 Membrane Filters

Unlike depth filters, membrane filters commonly contain a well-defined pore structure and a consistent porosity range. These filters can remove all bacteria, yeast, and colloidal particles. Membrane filters can be used as dead-end filtration or tangential flow mode, and the filtration obtained by using such membrane filters is usually referred to as microfiltration. Either evaporation, stretching, quenching, or track-etched process produces membrane filters. In addition, these filters can be developed in various structures to meet the specific application. For instance, solutions containing high contamination loads can be filtered using asymmetric membrane filters. In this particular type of filter, the size of the pore on the upstream side is larger than on the downstream side of the membrane. In aseptic processing to remove bacteria from liquids/gases, membrane filters having a rated pore size of 0.2 μm are used as the classic sterilizing grade filters (Levy & Jornitz, 2006). It is to be noted that due to their well-defined pore structure and consistency in porosity range, polycarbonate track-etched membrane filters are used in industry for reducing the size of liposomes and emulsions. A schematic of dead-end and crossflow filtration is shown in Fig. 9.2.

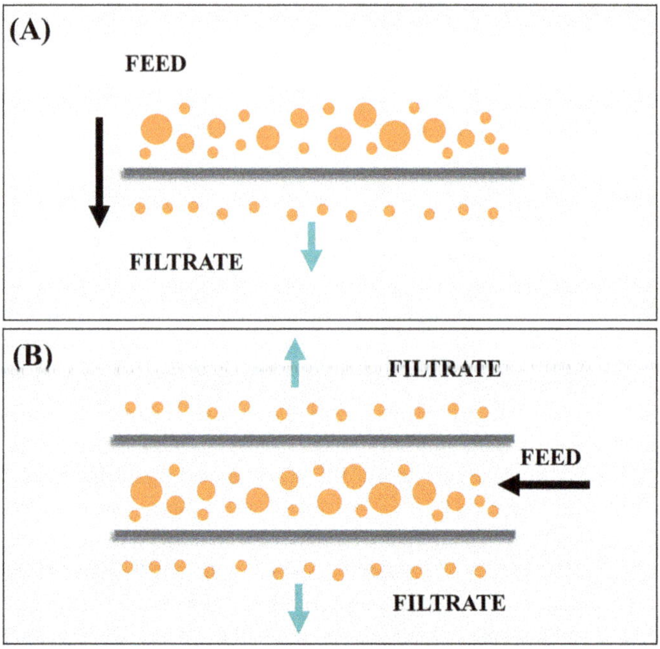

Fig. 9.2 Schematic of (**a**) Dead-end filtration; and (**b**) Crossflow filtration

9.5.1.3 Reverse Osmosis Membrane

Reverse membranes are widely used to treat water to remove ionic contamination. These membranes have an extremely small pore size, and extensive pressure is used on the upstream side of the filter membrane to force the liquids through the pores.

9.5.1.4 Ultrafilter Membrane

The retention ratings of ultrafilter membranes are expressed in molecular weight cut-off (MWCO) (i.e., the molecular weight of the drug substance to be retained). Usually, the range is 1000–300,000 Da. Ultrafiltration membrane filters are often used in cross-flow (tangential flow) systems. Ultrafiltration systems are generally used for processes like fractionation, concentration, and diafiltration steps of proteins, peptides, or viral vectors by enabling the removal of contaminants, buffer exchange, and concentration of a target protein without adversely stressing (such as shear forces) the target drug substance (Levy & Jornitz, 2006).

9.5.1.5 Nanofilter Membrane

Nanofilter membranes, which use size exclusion as the primary retention mechanism, are used to separate viruses, and the retention ratings of 20 and 50 nm are most commonly used (Levy & Jornitz, 2006).

9.6 Rating of Membrane Filter

Membrane filters can be rated either as nominal or as absolute. Nominal rating is based on the removal of particles or bacteria of a particular size in terms of percent weight basis. Therefore, when a filter manufacturer claims that their particular filter has a nominal rating of 75% for 0.2 μm or greater, it will remove 75% of all particles or bacteria ≥0.2 μm. On the other hand, the absolute rating defines the diameter of the largest particle that will pass through the filter. This means that a filter having an absolute rating of 0.2 μm means that no particle or bacteria greater than 0.2 μm will pass through such filter. It is to be noted that filters having an absolute rating of 0.2 μm are used in the industry for sterile filtration unit operation.

9.7 Applications of Sterilizing Grade Membrane Filters

Sterilizing grade membrane filters are widely used in the pharmaceutical and biopharmaceutical industries, and a few major applications are given below:

(a) Sterile filtration of small molecules that cannot tolerate heat or other terminal sterilization methods
(b) Sterile filtration of proteins and peptide-based biopharmaceuticals
(c) Sterile filtration of cell culture media used for the production of biopharmaceuticals
(d) Sterile filtration of buffer solutions and gases (including fermenter inlet air) used in the production of biopharmaceuticals

9.8 Membrane Polymers

The following section provides a summary of several types of membrane polymers used in filters (Reif, 2006):

9.8.1 Hydrocarbon-Based Polymers

These polymers are synthesized from vinyl monomers ($H_2C{=}CHR$). This category includes polypropylene membranes, which are both strong and versatile. However, the membranes have limited porosity and are mainly symmetric in nature. Being hydrophobic in nature, it is not suitable for the filtration of aqueous solutions (unless surfactants are included in the solution) and is reserved for the filtration of organic solvents. It is compatible with acidic and caustic solutions and most organic solvents, suggesting a broad range of applications. However, softening at around 80 °C was observed, so steam sterilization (121–134 °C) remains a concern and is a major limitation for usage. Radiation sterilization also results in autocatalytic degradation of these polymers, though the same can be prevented by the inclusion of additives.

9.8.2 Polyamides

Polyamides, known as "nylons", remain one of the most widely used polymers for manufacturing microfiltration membranes. Compared to aliphatic polyamide-based membranes, aromatic polyamide membranes are preferred specifically for the filtration of organic solvents as they have excellent chemical compatibility. On the other hand, the low cost and long usage history of aliphatic polymer membranes make them very common for microfiltration-based applications. Aliphatic polyamide-based membranes are relatively hydrophilic in nature due to the presence of weak charges on the polymer backbone. The membrane has very high adsorption characteristics, and hence, the filtration of high-potency drugs remains a concern.

9.8.3 Polysulfone

Sulfone-containing polymers are among the most important groups of polymers for membrane filters. All polysulfone-based polymers used commercially are almost amorphous in nature. They are relatively polar, and despite this fact, they adsorb very little water and, hence, exhibit no noticeable swelling in an aqueous solution. In addition, polymers have extremely high resistance to hydrolysis across the entire pH range, even at steam sterilization temperatures. The membrane matrix is very robust, stable against ionizing radiation, as well as heat stability up to >200 °C. Polysulfone and polyethersulfone are the two most commonly used commercial membrane polymers. These membranes can be manufactured either highly symmetric or asymmetric or even a combination of both. The very high porosity of the membrane imparts excellent filtration rates. In addition, the adsorption of the proteins, peptides, and preservative is very low. Therefore, microfiltration membranes made of these two polymers are preferred for the filtration of biopharmaceutical drug products.

9.8.4 Fluorpolymers

Among the various fluorpolymers, poly(vinylidene fluoride) (PVDF) and poly (tetrafluorethylene) (PTFE) are the most commonly used polymers for the manufacturing of membranes. These polymers have high chemical, oxidative, and thermal stability, as well as excellent compatibility with most solvents. However, these polymers cannot withstand radiation sterilization. Hydrophobic PTFE membrane remains the material of choice for the filtration of chemicals and hot air as it has very high thermal stability (>260 °C) and resistivity against almost all solvents.

In comparison to PTEE, PVDF is less hydrophobic, and hence, it is not suitable for air filtration. In addition, to make it suitable for the filtration of water-based solutions, the surface of the PVDF membrane requires treatment with a hydrophilic material like acrylic acid. However, this surface treatment reduces the chemical stability of the whole membrane. Therefore, PVDF membranes find limited applications in microfiltration.

9.8.5 Cellulosic Polymers

Cellulose acetate remains one of the most common cellulose based materials used in the manufacturing of microfiltration membranes. Cellulose acetate-based membranes are hydrophilic as well as stable against weak basic and acidic solvents. The membrane also has stability against high temperatures as well as physical stress. In addition, the extremely low non-specific adsorption of chemical species, including proteins and peptides, by cellulose acetate membrane makes this material preferred choice for sterile filtration of potent drug solutions, in which the non-specific adsorption-related drug/preservative loss due to the filter-membrane remains a major concern.

9.8.6 Polycarbonates

Polycarbonate polymer has limited chemical compatibility with strong acids and most organic solvents; however, a combination of unique features, including superior toughness, high resistance to heat, lower price, higher transparency, and acceptable compatibility with aqueous and aliphatic solvents as well as alcohol-based solvents make this polymer popular in device construction. In the polycarbonate-based polymer, bisphenol A polycarbonate is the most economical; however, when compared to other existing polymers as a membrane base polymer, polycarbonates have very low porosity, especially when manufactured using conventional techniques. Interestingly, the track-etching process that enables the production of membranes with symmetric pores requires polymer, which is very robust and tough. Certainly, polycarbonate is the polymer that meets this requirement, and therefore, most membranes are made up of this polymer.

9.9 Filter Design and Construction

Flow rate and total throughput are the two most important parameters for a sterilizing grade 0.2 μm membrane filter. Flow rate becomes a primary target whenever the liquid is filtered with limited contaminants or fouling components. This is particularly important for the filtration of large volumes of fluid, such as buffers or large-volume parenteral, in which the completion of filtration in the shortest time would make the equipment available for reuse. It is to be noted that the flow rate depends on the whole filter cartridge design and not just the porosity, thickness, and construction design of the filter membrane. For example, if a membrane having an exceptional flow rate cannot be pleated, then such a membrane is of no use for the construction of a filter cartridge. Total throughput, a widely used performance criterion in most filtration applications, refers to the total volume of liquid that can be filtered before the filter blocks. Total throughput is directly proportional to the filter design, surface area, system size, and prefilter combinations, and it has a significant impact on filtration costs. The following section discusses various filter designs available for commercial use.

9.9.1 Disc Filter

Disc (flat) filters, specifically 293-mm discs, were the first filter configuration used in the industry. To increase the area available for filtration, many membrane discs were assembled in a multi-stack stainless steel filter housing. However, major limitations, including the extreme difficulty of housing wetted filters and wrinkles during assembly, led to replacing these filter configurations with pleated filter cartridge formats. Standard sizes in terms of diameter are 4, 25, 47, 50, 90, 142, and 293 mm, with 47 mm being the most common. Membrane discs of 47- and 50-mm print with coloured grids are utilized as microbial (analytical) assessment filters to facilitate the counting of bacteria in a defined filtration area and thereby the volume filtered previously. These analytical filters generally have a pore size of 0.45 μm. They are generally made up of Nylon or Cellulose Nitrate materials to capture the bacteria through an adsorption mechanism rather than sieve retention, as the sole mechanism seen with 0.2 μm membranes (Jornitz, 2006a).

9.9.2 *Cartridge Filters*

Unlike disc or flat filters, the unique feature of a pleated cartridge filter is the zigzag or accordion shape folding of the membrane with upper and lower support layers to enable an extensively large effective filtration area. To replace a ten-inch filter cartridge and achieve the same filtration area, 15 numbers of 293 mm discs would be needed (Jornitz, 2006a). Not only this, even the floor space requirement for such 15 discs would be very high in comparison to a small space requirement for a single cartridge and providing the same effective filtration area. Further, each disc filter requires O-ring sealing, confirming that disc filter assembly is highly time-consuming. A significant increase in effective filtration area with the pleated cartridge filter helped drug product manufacturers in terms of requirement of lower applied differential pressures and hence enabled larger volume flows, specifically for sterile filtration of viscous liquids.

9.9.3 *Capsule Filters*

Disk and cartridge filters are disposable, while their housings and holders, made of metal, are non-disposable (Jornitz, 2006a). However, capsule filters are encapsulated in a plastic housing, and the whole unit is disposable. This kind of filter format has some unique advantages. For example, capsule-based filters are available as non-sterile as well as sterilized. Therefore, whenever sterile filters are needed, they are readily available. In addition, disposable capsule filters eliminate the risk of cross-contamination. On the contrary, a cartridge filter has to sterilized after assembling and then only it can be used for sterile filtration. It is important to note that data shows that when the total cost of a cartridge filter, including labour costs involved in assembling it, was compared to that of a disposable capsule filter, it was later not found to be expensive.

9.10 Filter Validation

It is impossible with currently available technology to measure the sterility of each filled container. Therefore, sterility assurance of the filtered product can only be assured through validation of the filtration process. For this reason, validating a sterilizing grade filter is highly critical. A guidance document published by the United States Food and Drug Administration (US FDA) and European Medicines Agency (EMA) discusses the requirements for sterilizing filters and their validation (EMA, 2019; U.S. FDA, 2004).

Validation of the sterile filtration process generally takes into consideration three things: evaluating the effect of the pharmaceutical liquid product on the filter, evaluating the effect of filter on the filtered product, and proving that the filter is capable of removing all bacteria from the liquid under actual filtration conditions, thereby resulting in a sterile filtrate (Madsen, 2006). These three things are evaluated by performing various tests, which can be divided into two broad categories: destructive testing and non-destructive testing. First category of tests is used initially to qualify the filter before it can be put into usage for sterilization of routine batches of a drug product. On the other hand, test belonging to the second category is performed before and after the usage of a filter, whenever the same is used for sterilization of routine production batches of a particular drug product. Destructive testing includes three main tests; (a) bacterial retention using the actual final drug product; (b) filter extractable and leachable; and (c) compatibility of the filter with the actual drug product. It is to be noted that filter validation study is outsourced by pharmaceutical industries. Once a specific filter for a particular drug product has been identified by a formulation scientist, then the manufacturer of that specific filter will perform the filter validation study. Although filter validation study is done by filter manufacturer, it is the responsibility of the drug product manufacturer that the study complies with the regulatory requirements of a particular regulatory agency.

9.11 Bacterial Retention Test

In the bacterial retention test, the filter is challenged with a general population of *Brevundimonas diminuta* (formerly *Pseudomonas diminuta*) in the actual drug product. This bacterium is used as it has a size of approximately 0.3 μm, barely above the absolute rating of the 0.2 μm sterilizing grade filter. Previously, it was allowed to use a placebo formulation having the same critical attributes like osmolality, pH, viscosity, surface tension, and ionic strength; however, now it is mandatory to use the actual drug product. If the formulation has intrinsic antimicrobial activity, it must first be neutralized before adding the challenge bacteria. The concentration of bacterial cells must be 10^7 cells per cm^2 filter surface area. Various processing conditions which are generally simulated include usage of the same filter type and configuration (disc or cartridge) that will be used in the filtration of the actual drug product, filtration pressure and flow rate, duration of the filtration process, and temperature (Akers, 2010).

9.12 Filter Integrity Testing

Sterilizing-grade membrane filters are to be tested for integrity to ensure that the filters are integral and fulfil the purpose. These tests are called filter integrity tests and may be performed before. However, it must be performed after the filtration process (Jornitz, 2006b). It is to be noted that this test is not only applicable for filters used for liquids, but it is also applicable for air filters. There are various filter integrity tests, such as bubble point, pressure hold, multipoint diffusion test, and water intrusion test. The bubble point test is usually performed on smaller filters, while the diffusion test is generally used for filters with large surface areas. This is because as the surface area of filters becomes large, diffusion of air through the water-filled pores tends to obscure the bubble point. Similarly, the pressure hold test also can be applied to large surface area filters. It is the filter manufacturer who will recommend the best integrity test for a particular filter system.

9.13 Conclusion

In this chapter, we have briefly discussed various types of mixtures such as positive mixture, negative mixture, as well as neutral mixture. Various aspects of solid-liquid mixing, in particular suspension was discussed along with various equipment used in their manufacturing. Various types of filters used in the pharmaceutical industries were also discussed. Sterility assurance of pharmaceutical products sterilized by filtration sterilization depends solely on the filter validation and this aspect of sterilizing grade membrane filter was also discussed in the chapter. Sterilizing grade filters made up of different membrane polymers are developed and available for commercial usage. All the polymers are equally capable of producing the sterile filtrate; however, the one which has least effect on formulation components such as minimal adsorption of preservative, peptide, or protein drugs should be considered. This aspect of various membrane polymers is also discussed in the book chapter.

References

Atiemo-Obeng, V. A., & Calabrese, R. V. (2003). Rotor–Stator mixing devices. In E. L. Paul, V. A. Atiemo-Obong, & S. M. Kresta (Eds.), *Handbook of industrial mixing* (pp. 479–505). Wiley.

Dickey, D. S. (2015). Fluid mixing equipment design. In P. J. Cullen, R. J. Romañach, N. Abatzoglou, & C. D. Rielly (Eds.), *Pharmaceutical blending and mixing* (pp. 311–344). Wiley.

Etchells, A. W. (2001). *Mixing of floating solids*. InPlenary Conf. ISMIP4.

European Medicines Agency. (2019). *Guideline on the sterilisation of the medicinal product, active substance, excipient and primary container*. EMA.

Hemrajani, R. R., & Tatterson, G. B. (2003). Mechanically stirred vessels. In E. L. Paul, V. A. Atiemo-Obong, & S. M. Kresta (Eds.), *Handbook of industrial mixing* (pp. 345–389). Wiley.

Jornitz, M. W. (2006a). Filter construction and design. In T. Scheper (Ed.), *Sterile filtration. Advances in biochemical engineering/biotechnology 98* (pp. 105–123). Springer.

Jornitz, M. W. (2006b). Integrity testing. In T. Scheper (Ed.), *Sterile filtration. Advances in biochemical engineering/biotechnology 98* (pp. 143–180). Springer.

Karbstein, H., & Schubert, H. (1995). Developments in the continuous mechanical production of oil-in-water macro-emulsions. *Chemical Engineering and Processing Process Intensification, 34*(3), 205–211.

Kolhe, P., Shah, M., & Rathore, N. (Eds.). (2013). *Sterile product development: Formulation, process, quality and regulatory considerations*. Springer.

Levy, R. V., & Jornitz, M. W. (2006). Types of filtration. In T. Scheper (Ed.), *Sterile filtration. Advances in Biochemical Engineering/Biotechnology 98* (pp. 1–26). Springer.

Madsen, R. E. (2006). Filter validation. In T. Scheper (Ed.), *Sterile filtration. Advances in biochemical engineering/biotechnology 98* (pp. 125–141). Springer.

Morales, J. O., Watts, A. B., & McConville, J. T. (2016). Mechanical particle-size reduction techniques. In R. O. Williams III, A. B. Watts, & D. A. Miller (Eds.), *Formulating poorly water soluble drugs. AAPS Advances in the Pharmaceutical Sciences Series 22* (pp. 165–213). Springer.

Özcan-Taşkin, G. N. (2015). Dispersion of fine powders in liquids: Particle incorporation and size reduction. In P. J. Cullen, R. J. Romañach, N. Abatzoglou, & C. D. Rielly (Eds.), *Pharmaceutical blending and mixing* (pp. 129–151). Wiley.

Pacek, A. W. (2015). Emulsions. In P. J. Cullen, R. J. Romañach, N. Abatzoglou, & C. D. Rielly (Eds.), *Pharmaceutical blending and mixing* (pp. 183–232). Wiley.

Reif, O. W. (2006). Microfiltration membranes: Characteristics and manufacturing. In T. Scheper (Ed.), *Sterile filtration. Advances in Biochemical Engineering/Biotechnology 98* (pp. 73–103). Springer.

Sterile Filtration. (2010). In M. J. Akers (Ed.), *Sterile drug products formulation, packaging, manufacturing and quality* (pp. 267–277). Informa Healthcare.

Twitchell, A. M. (2012). Mixing. In K. Taylor & M. E. Aulton (Eds.), *Aulton's pharmaceutics: The design and manufacture of medicines* (p. 156). Elsevier.

U.S. Food and Drug Administration. (2004). *Guidance for industry sterile drug products produced by aseptic processing – Current good manufacturing practice*. FDA.

Utomo, A., Baker, M., & Pacek, A. W. (2009). The effect of stator geometry on the flow pattern and energy dissipation rate in a rotor–stator mixer. *Chemical Engineering Research and Design, 87*(4), 533–542.

Chapter 10
Scale-Up of Liquid Mixing Process

Kedar S. Prayag and Anil B. Jindal

Contents

Abstract To meet the growing demands of production and reduce costs, it is necessary to increase the scale of production. However, transferring the lab-scale process to a large manufacturing scale can be a challenging task for trained personnel. The manufacturing process, which involves homogeneous or heterogeneous systems, requires several scale-up parameters to be considered when designing large-scale production. These parameters include power requirements, volume measurement, vessel design, mixing device speed, and any other equipment that affects the flow and mixing quality. In this chapter, we will discuss the basics of fluid motion, factors that affect it, mixing devices, and auxiliary equipment that assists liquid mixing. We will also provide an overview of the various approaches to scientific scale-up, including relevant equations and an illustrative example.

Keywords Scale-up · Liquid mixing · Impeller · Baffles · Flow pattern · Agitation

---

K. S. Prayag · A. B. Jindal (✉)
Department of Pharmacy, Birla Institute of Technology and Science, Pilani, Jhunjhunu, Rajasthan, India
e-mail: anil.jindal@pilani.bits-pilani.ac.in

A. B. Jindal (ed.), *Pharmaceutical Process Engineering and Scale-up Principles*, AAPS Introductions in the Pharmaceutical Sciences 13,
https://doi.org/10.1007/978-3-031-31380-6_10

10.1 Introduction

Liquid mixing is the process of combining two or more different liquids using mechanical energy to achieve random distribution. This mechanical energy creates an agitation pattern that promotes mixing in the fluids and leads to specific flow patterns, such as circulation, which brings about homogeneity in the mixture. Mixing is important in many industries, such as food and beverage, pharmaceuticals, and chemical manufacturing, to ensure that products are uniform and meet quality standards.

Mixing fluids with or without solids in the system is thought to depend upon the natural or forced convection during the mixing process (Block, 2005). Fluid motion can be aided by pulses or vibrations, which create turbulence and move material in new, random paths. Mixing helps to increase homogeneity and achieve the desired operational outcome. There are various methods of mixing liquids, including mechanically agitated vessels, rotor-stator mixers, and in-line static mixers. Each method has its advantages and disadvantages, and the most suitable method will depend on the specific requirements of the application (The 6 most common fluid mixing processes, 2022). Mechanically agitated vessels are among the most commonly used mixing devices. They work to reduce inhomogeneity and nonuniformity in the system by achieving the desired output from the mixed fluid system through mixing. These vessels typically use an impeller or other mechanical means to agitate the liquid and promote mixing. They are suitable for a wide range of applications and can handle a variety of fluids but may not be as efficient as other types of mixers in certain situations (Gates & Henley, 1985).

Three main principles govern the flow properties and mixing behavior of fluids: the law of mass conservation, the law of energy conservation, and the fundamental laws of motion. The challenges in scaling up liquid mixing can be attributed to the complex behavior of fluids governed by these principles (Am Ende & Am Ende, 2019). Due to a lack of understanding of the mechanisms involved in liquid agitation and blending, it can be difficult for researchers to develop an efficient scale-up process from the lab to production scale. To overcome this challenge, it is important to understand the mixing efficiency and the precise mechanisms of material transfer and fluid motion, as well as how they are interconnected. The fluid dynamics, design, and geometry of mixing vessels, impellers, baffles, and power consumption requirements across scales are important factors to consider when scaling up liquid mixing processes (Rushton, 2002). This chapter covers various aspects of the liquid mixing process, including the mechanisms involved and different approaches that can be taken when scaling up the process.

10.2 Mechanism of Liquid Mixing

(a) *Bulk transport*: Bulk transport refers to the movement of a large amount of material from one location to another. This can involve convection, where the material is moved with the help of paddles or impellers in a vessel. However, simple material circulation may not always result in adequate mixing (*Mixing: Mechanisms of Mixing.*, 2021).
(b) *Turbulent mixing*: Turbulent mixing is a type of mixing that occurs when a fluid is agitated or moved in an irregular or random pattern. This can be caused using impellers, turbines, or other types of agitators that create flow patterns within the fluid. The pressure within the fluid can fluctuate as the impeller rotates and creates eddies, or small swirling patterns, within the fluid. These eddies help to mix the fluid homogeneously and can be used to mix different components within a vessel effectively (Lévêque, 2006). Eddies represent the portion of fluid moving in any direction, often irrespective of the fluid flow in the vessel.
(c) *Laminar mixing*: Laminar mixing is a type of mixing that occurs when the fluid adjacent to the agitator moves in a parallel pattern to the flow direction. This type of mixing is characterized by forming laminar or streamlined, flow patterns within the fluid. Laminar mixing is typically used for viscous fluids, as the shear stress within the fluid is directly proportional to the strain applied across the layers of fluid.
(d) *Molecular diffusion*: Molecular diffusion is a type of mixing that occurs at the microscopic level, as the movement of individual molecules from a region of higher concentration to a region of lower concentration leads to the mixing of different substances. This process continues until equilibrium is reached, at which point the concentration of the substances is uniform throughout the mixture. Molecular diffusion is a relatively slow process, but it can still be an important mechanism for mixing on the molecular scale, especially in situations where other types of mixing are not possible or practical.

10.3 Important Considerations During Liquid Mixing and Its Scale-Up

10.3.1 Flow Pattern

The fluid flow pattern during mixing can depend on various factors, including the type of mixing device and associated baffles used, the properties of the fluid being mixed, and the speed and geometry of the mixing impeller. The Reynolds number (R_e) is a dimensionless number that is commonly used to describe the flow pattern of a fluid and to predict the behavior of the fluid under different mixing conditions. The Reynolds number is defined as the ratio of inertial forces to viscous forces in a fluid and is given by the equation

$$R_e = \frac{\rho V D}{\mu} \quad (10.1)$$

where ρ is the density of the liquid, V is the velocity of flow, D is the diameter of the pipe, and μ is the viscosity of the fluid under mixing operation. If the value of $R_e > 2000$, the flow is considered turbulent, whereas $R_e < 2000$ is regarded as laminar flow, wherein the values in between exhibit transitional flow patterns. Furthermore, the concept of Reynolds number can be extended in stirred tanks with impellers, denoted by the following equation where N is the RPM of the impeller.

$$R_{\mathrm{eimp}} = \frac{\rho N D^2}{\mu} \quad (10.2)$$

The value of $R_{\mathrm{eimp}} < 10$ is considered to have a laminar flow pattern, $R_{\mathrm{eimp}} > 10^4$ has a turbulent flow regime, whereas values lying between 10 to 10^4 exhibit transitional flow patterns.

Flow patterns can be classified as either laminar or turbulent, depending on the value of the Reynolds number. In laminar flow, the fluid moves in a smooth, orderly pattern and is characterized by low Reynolds numbers. In turbulent flow, the fluid moves in an irregular, chaotic pattern and is characterized by high Reynolds numbers. The transition between laminar and turbulent flow can be influenced by various factors, including the type of mixing device, the properties of the fluid, and the geometry of the mixing vessel.

High-viscosity fluids represent most non-Newtonian fluids, and because of their complicated fluid rheology, mixing characteristics differ significantly from routine Newtonian fluids. To address such issues, the power number P_o is a dimensionless variable widely used as a function of the Reynolds number.

10.3.2 Rheology of the Fluids

It is important to study the rheology of fluids during the mixing process because the viscosity and flow properties of a fluid can have a significant impact on the efficiency and effectiveness of the mixing process. Rheology is the study of the deformation of fluids under the influence of stress and strain, and it is an important factor in many industrial operations that involve the mixing of fluids with complex rheological properties. Fluids can be classified as thixotropic, shear-thinning, or viscoelastic based on their rheological properties. Thixotropic fluids exhibit a decrease in viscosity when subjected to shear stress, while shear-thinning fluids exhibit a decrease in viscosity with increasing shear rate. Viscoelastic fluids exhibit both viscous and elastic behavior and can be affected by both stress and strain. Mixing equipment must be carefully designed to ensure accurate mixing of fluids with these complex rheological properties, as the viscosity and flow characteristics of the fluid can vary significantly within the mixing vessel (Bossler et al., 2017).

Several techniques can be used to characterize the flow patterns and mixing time in stirred tanks, depending on the rheological properties of the fluid being mixed. Classic characterization techniques such as rheometry can be used to measure the viscosity and other rheological properties of the fluid. In contrast, flow visualization techniques such as Particle Image Velocimetry (PIV) and Planar Laser Induced Fluorescence (PLIF) can be used to visualize the flow patterns within the mixing vessel (Abadi, 2016).

10.3.3 Design of Liquid Mixing Devices

Impellers are an important component of mixing equipment, as the type, design, and rotation of the impeller can have a significant impact on the flow patterns and mixing performance of the system. Different types of impellers can be used to achieve different flow patterns and fluid pumping rates, and the design and arrangement of the blades can be modified to optimize the mixing process for different fluids and viscosities (Fig. 10.1).

Impellers can be broadly classified based on their flow and shear characteristics, and the force applied to the fluid during mixing can result in flow and shear. The most extreme turbulent intensity during mixing tends to occur near the impeller, and a large portion of the energy used in the mixing process is dissipated in this region. Turbulent vortices and fluctuations in the fluid can also be observed behind the impeller blades. The choice of the impeller and the mixing equipment design can be optimized to achieve the desired mixing behavior and performance for a particular application (Table 10.1).

Impellers can be broadly classified into three main categories based on the flow pattern they produce: tangential flow impellers, axial flow impellers, and radial flow impellers (Peters, 2015). Tangential flow impellers produce a swirling flow pattern that can be used for mixing fluids with low viscosity or for suspending solids. These impellers are typically used in applications where a high degree of mixing is required, such as in the production of emulsions or suspensions.

Axial flow impellers produce an up-and-down flow pattern and are often used in applications where the even distribution or suspension of solids is difficult. These impellers are effective at pumping fluids in a downward direction, which can be useful for preventing material settling during solid-liquid mixing. Examples of axial flow impellers include marine propellers, pitched blade turbines, and hydrofoils.

Radial flow impellers produce a flow pattern that is perpendicular to the shaft of the impeller and is often used for mixing fluids with high viscosity or for applications where a high degree of flow circulation is desired. These impellers can effectively provide good cross-mixing between different components within the fluid. In an experimental set-up, the positioning of the impeller does play a significant role in producing the flow patterns in the tank. For instance, the flow pattern created by a typical axial flow impeller makes an excellent top-to-bottom motion when the

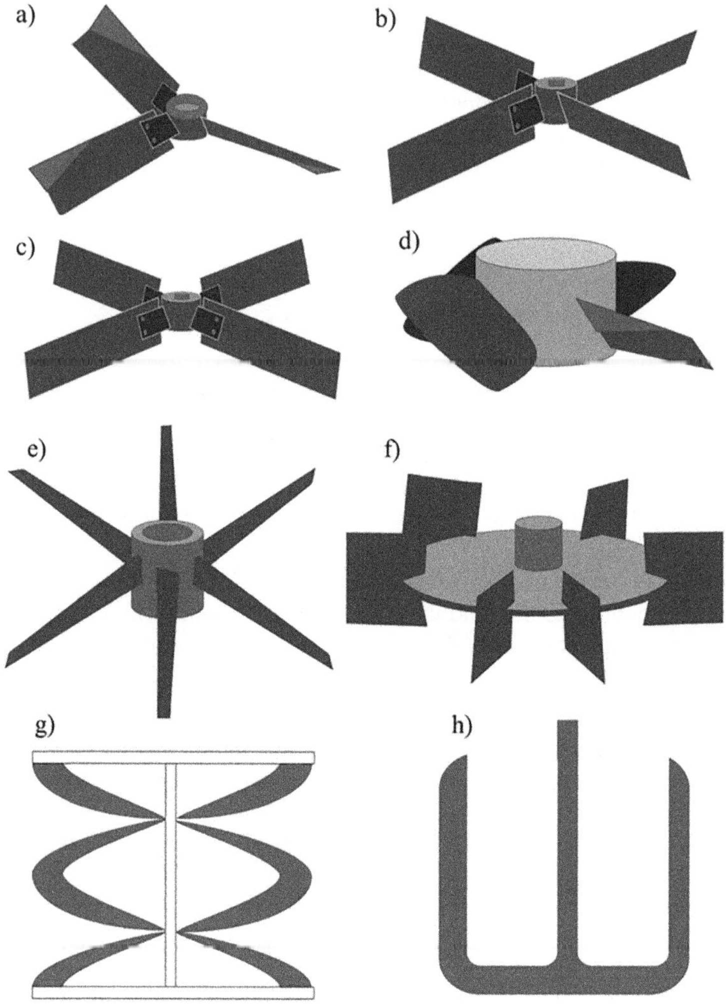

Fig. 10.1 Various types of impellers widely used in the manufacturing industry on the basis of their utility. (**a**) Narrow-blade, (**b**) Pitched-blade, (**c**) Straight-blade, (**d**) Wide-blade, (**e**) Narrow-blade turbine, (**f**) Flat-blade disc, (**g**) Helical ribbon, and (**h**) Anchor-blade impeller

agitator is center-mounted, and the vessel is baffled. These impellers commonly have four or 6-blade configurations. Whereas radial flow impellers do not produce a high tank turnover flow, they are known to attain higher shear and less flow after applying horsepower than the axial flow impellers.

Instead of pushing the fluid vertically away from the impeller, radial flow impellers tend to push it radially outward as a centrifugal force. High-shear impellers, such as flat-blade turbines or paddles, often have a radial flow design and are effective at providing good mixing and achieving a high degree of uniformity. The

Table 10.1 Types of impellers and their applications

Sr no	Impeller type	Applications
1.	Straight-blade turbine	Solid suspension Local liquid motion
2.	Wide-blade	Solid suspension Transitional flow
3.	Pitched-blade	Blending Dispersion
4.	Narrow-blade	Turbulent heat transfer Solid suspension
5.	Narrow-blade turbine	Turbulent blending Solid suspension
6.	Flat-blade disc turbine	Liquid-liquid dispersion
7.	Anchor	Viscous fluid mixing
8.	Helical ribbon	Viscous mixing
9.	Side-mounted wide blade	Wastewater circulation Oil storage

configuration of the impeller blades can have a significant impact on the performance of a mechanically agitated mixer, and the design of the impeller can be optimized to improve mixing quality and achieve the desired result (Torotwa & Ji, 2018).

Computational fluid dynamics (CFD) is a powerful tool that can be used to study the mixing performance of different impeller designs in stirred vessels. By using CFD, researchers can simulate the flow patterns and mixing behavior of fluids under different conditions and predict the performance of different impeller designs. This can be useful for reducing operational costs and minimizing experimental failures at the bench scale, as it allows researchers to study the mixing performance of different impellers without the need for costly and time-consuming experiments. In their study, Torotwa et al. used CFD to study the mixing performance of different impeller designs, including anchor, saw-tooth, counter-flow, and Rushton turbine impellers. They measured the electrical conductivity induced by the mixing of potassium sulfate granules in the mixing vessel to determine the extent of dissolved solid concentrations impacted by the rotation of the impeller. The computational models were developed using commercial software such as ANSYS Fluent 18.1 solver and the standard k-epsilon (ε) turbulence model (Torotwa & Ji, 2018).

10.3.4 Mixing Baffles

In a mixing tank, flat plates that cover the entire height of the tank and, in some instances, from the base of the tank are called as mixing baffles. They are auxiliary devices that purposefully have been kept between the walls of the tank and the impeller to guide the flow of fluid agitated during rotation of the impeller (Lu et al.,

1997). Baffles are long, flat plates mounted inside the tank by fastening using bolts coated with polyethylene and sealed with a gasket to stop leaks during mixing operation (PolyProcessing Solutions, 2022).

Mixing baffles have proved to be of extreme importance in stirred tanks to avoid fluid swirling or vertexing, thereby reducing the power consumption in the process and ultimately improving mixing efficiency. Furthermore, mixing baffles increase the turbulence and cross-mixing and provides an even distribution of kinetic energy dissipated by the agitator in the system, especially in the case of low-viscosity fluids (viscosity$<$5000 cP).

The major limitation in mixing baffles is that it can only be used in high-speed impeller applications and cannot be used in anchor or helix-shaped impellers with very close clearance with the tank. Moreover, baffles are also not suitable in case of mixing of solids that need wetting for their dissolution since baffles reduce vertexing significantly in the mixing process. The reduced vortex or eliminated vortex by baffles increases the mixing task challenge by many fold.

10.3.5 Airjet/Jet Devices

Airjet devices pass highly pressurized gases through nozzles, usually placed at the bottom of the tank. The jets are arranged so that the bubbles created during the passing of gases create buoyancy, lifting the fluid from the bottom of the tank to all over the tank wherever the gas spreads. The foremost important consideration in using an airjet system is that the fluid under mixing should be low viscous, non-foaming, and unreactive to the gas utilized.

Similarly, when the jets of fluid to be mixed are pumped into the system through the nozzle, it creates turbulence in the tank filled with fluid resulting in proper circulation of the fluids. A low-shear rotor-stator mixing head is used by the jet mixer, which creates a unique tank flow pattern that effectively and consistently keeps solids in suspension or evenly blends the liquids in a vessel (Bathija, 2022). Unlike, traditional impeller systems, jet mixers do not produce liquid vortex, reducing air entrapment and subsequent oxidation of the fluids under the mixing process (Transvac, 2021). Jet mixers find immense importance in various mixing processes when introduced in vessels, storage tanks, and neutralization basins.

10.3.6 Heat Transfer

During the mixing process, physical contact between two or more liquids can result in the exchange of thermodynamic parameters, including heat and entropy. The temperature difference between the liquids, the heat generated by the mixing impellers, and the resistance created by the baffles used in the vessel can all contribute to heat generation during the mixing process. This can be particularly important to consider in situations where the mixed liquids contain thermosensitive actives, viscoelastic fluids, or non-Newtonian fluids, as these fluids can be sensitive to changes in temperature or the application of shear stress. In these cases, it may be necessary to study the heat exchange process to optimize the mixing process and prevent damage to the active ingredients or the integrity of the mixture.

In commercially available mixing equipment, specialized impellers and configurations are frequently used in mixing equipment for highly viscous fluids to reduce high shear zones and heat transfer (Luo et al., 2021). The Nusselt number (Nu) is a dimensionless number that is commonly used to describe the heat transfer phenomenon that occurs during the mixing of two or more liquids in a vessel. It is defined as the ratio of convection heat transfer to fluid conduction heat transfer and is given by the equation: Nu = (convection heat transfer) / (fluid conduction heat transfer). The Nusselt number represents the fraction of heat transmitted by fluid-to-fluid heat conduction to that transferred by forced convection with the help of impellers. To accurately study the heat transfer phenomenon during mixing, it is essential to maintain geometric similarity across scales (Hartmanshenn, 2020).

In several mixing operations involving non-Newtonian, viscous fluids, temperature maintenance is necessary to carry out adequate mixing. For instance, Borowski et al. investigated the heat transfer phenomenon in a viscous fluid by means of a multi-ribbon mixer (Heat transfer and temperature difference of 2 liquids, 2013).

$$Q = A \times U \times \blacktriangle T \quad (10.3)$$

where Q is heat exchange quantity, A is heat transfer area, U is heat transfer coefficient, and $\blacktriangle T$ is the temperature difference in two liquids under mixing.

Heat transfer during liquid mixing can have a significant impact on the scale-up of liquid formulations. At the laboratory scale, the surface area to volume ratio of the mixing tank is usually larger than at the pilot or production scale. This means that the heat transfer rate is generally faster at the small scale than at the large scale. Because heat transfer is a surface phenomenon, the amount of heat generated during the mixing process depends on the volume of the liquid present. As the scale increases, the amount of heat generated increases proportionally with the volume, while heat transfer remains proportional to the surface area. This can lead to heat accumulation inside the liquid, which can cause degradation of the active ingredients if they are thermolabile. To avoid heat accumulation inside the liquid during scale-up, jacketed vessels are often used at the pilot or production scale. In these vessels, cold water is continuously circulated on the outer surface of the mixing vessel, which increases the temperature difference and enhances the rate of heat transfer. Jacketed vessels are commonly used to mix liquids at the pilot or production scale to prevent heat accumulation and protect thermolabile active ingredients.

10.3.7 Material Transfer

Material or mass transfer is the movement of a small amount of one phase (usually a solute) to a larger amount of another phase (usually a solvent) in a mixing vessel. Liquid mass transfer typically occurs through convection, which involves the movement of molecules caused by the movement of impellers within the mixing equipment. The mechanism of liquid mixing can be described using Fick's first law of diffusion, which states that the movement of molecules occurs from a region of higher concentration to a region of lower concentration, which is given by Eq. 10.4.

$$\frac{Dm}{dt} = -D.A.\frac{dc}{dx} \qquad (10.4)$$

Diffusion coefficient, *D*, depends upon the size and viscosity of the diffusing molecules. Whereas *Dm*/*dt* represents the rate of mass transfer, and *dc*/*dx* represents the concentration gradient across boundaries. The concentration gradient is a decreasing function of time, which approaches zero at the end of the completion of the mixing process.

Furthermore, mass transfer coefficients play a significant role in conducting scale-up of gas-liquid scale-up. In such type of system, irrespective of geometric similarity, agitation rates, or any other mixing parameter, matching of mass transfer phenomenon is extremely important (*Mixing and Mass Transfer | Reaction Rate-METLER TOLEDO*, 2015).

10.4 Scale-Up of Liquid Mixing Process

The quality of the mixing process can be influenced by various factors, including the geometry of the tank, the shape and size of the impeller, the presence and arrangement of baffles, the density of the fluids being mixed, and the rheological properties of the fluids (such as viscosity and yield stress). When scaling up a mixing process from a lab or pilot scale to a larger production scale, it is often necessary to make trade-offs between cost and quality. It is essential to have a good understanding of the core mixing principles and the specific requirements of the process to achieve the desired outcomes and minimize unintended effects. There are various scale-up variables, formulas, and principles that can be used to establish a correlation between lab-scale and pilot-scale. Still, no single approach is effective in all circumstances. The most common approach for the liquid mixing scale-up is to keep any one of the factors, including shaft speed, power number, Reynolds number, Froude number, or tip speed constant across the scale (Kaalva, 2016). Moreover, fluid properties such as density, viscosity, fluid type, impeller type, tank size, and baffle arrangement may also affect the scale-up process.

10.4.1 Constant Power/Volume Function

Maintaining homogenous turbulence in the mixing vessel is important for ensuring a successful mixing process, and the rotational speed of the impeller can have a significant effect on the level of turbulence achieved. One way to determine the proper speed for maintaining homogenous turbulence across different scales is by considering the power consumption during the mixing operation. To scale-up a mixing process, it is necessary to predict the power consumption for the larger-scale mixing process and determine the optimum rotational speed. This can be done by multiplying the power consumption on a small scale by a scale-up factor. The scale-up factor considers the changes in the size and geometry of the mixing vessel, as well as the properties of the fluids being mixed, as the scale is increased.

The Kolmogorov's theory describes energy dissipation anticipated to the formation of eddies during mixing operation (Kuksin & Kuksin, 2021). In fluid dynamics, eddies are formed due to the swirling of a fluid and the reverse current produced when the fluid is in a turbulent flow state. The moving fluid generates a gap on the downstream side of the object where no downstream-flowing fluid is present. Theoretically, energy dissipation denoted by 'ε' describes the relationship between turbulence and mass transfer process during mixing, characteristic of eddies formed during the mixing process. Kolmogorov's theory explains the energy transfer phenomenon from larger to smaller eddies by energy dissipation phenomenon. It is the most widely used practical approach to scale-up in which maintaining a constant rate of turbulent energy dissipated (ε) as a function of P/V and expressed as

$$\varepsilon = \frac{P}{\rho V} \tag{10.5}$$

wherein ρ is the density of the liquid. Kolmogorov's length scales are essential determinants in predicting scale up of the mixing process formed due to eddies during mixing wherein the parameters like energy dissipated (ε) and kinematic viscosity (ν) can be expressed in terms of length (η), velocity (μ_η) and time (τ_η) scales such as

$$\eta = \left(\frac{\nu^3}{\varepsilon}\right)^{1/4} \tag{10.6}$$

$$\mu_\eta = (\varepsilon\nu)^{1/4} \tag{10.7}$$

$$\tau_\eta = \left(\frac{\nu}{\varepsilon}\right)^{1/2} \tag{10.8}$$

From Eq. 10.6, length scale represents that kinematic viscosity due to viscous forces are equivalent to inertial forces due to turbulent velocity fluctuations (Kim & Saito, 2013). This length scale parameter given by Kolmogorov suggests that the scaling factor for any mixing process can be dependent on ε, since kinematic viscosity (ν) is independent of scale merely keeping constancy in ε would result in equivalent η across any scale-up process. Use of ε or P/V for scale up purpose essentially requires maintenance of geometric similarity across scales. Furthermore, use of power number (N_P) a dimensionless number which is characteristic of impeller and vessel geometry is one of the parameters to be taken into consideration of scale-up process of mixing. For instance, scale up involving use of ε or P/V can be given by the use of various parameters including impeller diameter (D), rotational speed (N), volume of liquid (V), and density of liquid (ρ) as

$$\varepsilon = \frac{N_P N^3 D^5}{V} \tag{10.9}$$

$$\frac{P}{V} = \frac{N_P \rho N^3 D^5}{V} \tag{10.10}$$

Hence N_P can be given by Eq. 10.11, wherein the number of factors affects the power number such as number of blades, position of impeller in the vessel, number and dimensions of baffles in the vessel, etc. Finding N_P experimentally is also of immense importance via measuring torque on the impeller shaft, and measuring power under well-defined experimental conditions. N_P essentially signifies the drag coefficient or friction factor in the system represented by

$$N_P = \frac{P}{\rho N^3 D^5} \tag{10.11}$$

More importantly, turbulence created by the impeller given by power number provides necessary turbulence in the vessel under mixing operation. Further, fluids in a motion are usually described as laminar or turbulent depending on the velocity of the fluid quantified by another dimensionless parameter called Reynolds n12umber (N_{Re}) denoted as

$$N_{Re} = \frac{\rho D^2 N}{\mu} \tag{10.12}$$

Values of $N_{Re} < 100$ indicate laminar flow whereas, more than value of more than 10^4 signifies turbulent mixing in the system.

Further, calculating Froude number, another dimensionless number also signifies the power requirements across scale which is represented by the following equation

$$N_{Fr} = \frac{N^2 D}{g} \tag{10.13}$$

where N is the rotational speed, D is the diameter of the impeller, and g is the acceleration due to gravity exerted in the vessel (Prayag et al., 2021).

In a homogenously stirred tank, if all the types of similarities are fulfilled, including geometric kinematic, and dynamic, then a combined dimensionless equation can be represented by

$$N_P = K.N_{Re}^{-b}.N_{Fr}^{-d} \tag{10.14}$$

The constant K and exponents b and d need to be determined based on the type of agitator, its size and location in the tank, the dimensions of the tank, the depth of the liquid, etc. In order to achieve dynamic and kinematic similarities, prior fulfilment of geometric similarity is must. For instance, kinematic similarity requires similarity in ratios of velocities between corresponding points in one scale to be equal to those in the other scale, whereas dynamic similarity requires similarity in ratio of forces across scales.

10.4.2 Principle of Geometric Similarity

Designing a vessel, impeller, and baffles of the equipment holds enormous importance in scale-up of any liquid mixing process. The principle of geometric similarity involves multiple factors which affect the quality of scale-up process, including similarity of the dimensions of the equipments used across the scales, equal fluid velocity, equal surface motion, equal mass transfer rates, equal solid suspension, and equal liquid motion. In the case of liquid mixing, the different geometric dimensions, including the diameter of the tank (T), the diameter of the impeller (D), and the height of the liquid (Z) in the tank at different scales, are considered. To achieve similarity in liquid mixing during scale-up, the rotational speed is one of the significant determinants of achieving turbulence in the system. The prior determination of the geometric scaling factor (R) is essential. The ratio of D_1/T_1 of scale equipment or vessel is compared with a ratio of D_2/T_2 of large scale and calculated as Eq. 10.15.

$$R = \frac{D_1}{T_1} = \frac{D_2}{T_2} \tag{10.15}$$

where D_1 T_1 are the diameter of impeller and tank for small vessel and D_2 T_2 are diameters of impeller and tank for large vessel utilized for scale-up operation. Once geometric scaling factor (R) is determined, rotation speed required in scale-up batch can be obtained from Eq. 10.16, where N_2 and N_1 are rotational speeds in large-scale and small-scale batches, respectively (Fig. 10.2).

Further, 'η' describes the power law component wherein the values have been assigned from 0 to 1 in the literature concerning physical interpretation done earlier while designing the scale-up of the process. For instance, considering various parameters in scale-up such as equal blending time, surface motion, mass transfer rate, % of solids, and average fluid velocity has been assigned with values of 0, 1/2, 2/3, 3/4, and 1, respectively.

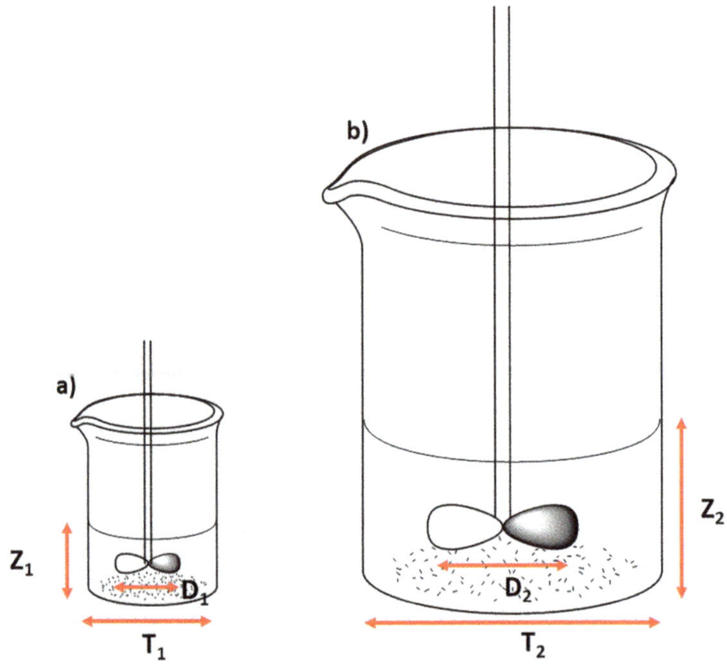

Fig. 10.2 Dimensions of vessels and apparatus used in (**a**) lab-scale batch, and (**b**) pilot-scale batch forms the basis of attaining geometric similarity across various scales

$$N_2 = N_1 \left(\frac{1}{R}\right)^{\eta} \tag{10.16}$$

In most scientific scale-up approaches, parameters such as D/T and Z/T ratios between 0.15 to 0.6 and 0.3 to 1.5, respectively, represent optimum mixing in the system efficiently.

Practically there are several limitations to this theory in practice. For instance, as per geometric rules, when vessel size doubles in diameter, volume increases by a factor of 8, which makes it challenging to achieve the volume/ diameter ratio required in geometric similarity.

10.4.3 Tip Speed Calculation

Technically, tip speed is tangential velocity at largest diameter of the impeller denoted by S_t and given by

$$S_t = \Pi.D.N \tag{10.17}$$

where D is the diameter of impeller and N is the rotation of the impeller. In case of geometrically scaled vessel with equal mean energy dissipation rate, tip speed remains higher in larger vessels bringing turbulence in the system regardless of the viscosity of the fluids under mixing.

10.4.4 Scale of Agitation Approach

The scale of agitation approach is based upon the equal liquid motion principle. According to the scale of agitation approach, it is attributed that equal liquid motion across the scale confirms similarity in the liquid mixing process. The fundamental principle behind the scale of agitation is achieving geometric similarity with a power-law component of $\eta = 1$, where a value of 1 denotes the minimal intensity required for the liquid to move completely. In contrast, a value of 10 denotes the greatest practicable intensity. Further, 1 or 2 represent mild agitation, between 3 and 6 denote medium intensity, while values between 7 and 10 indicate intense agitation in the system. In this approach, bulk fluid velocity (V_b) is calculated by considering several parameters, including terminal pumping number (N_Q), and adequate pumping capacity (Q).

The method is utilized mainly in vertical cylinders where the flow of liquids is primarily turbulent. It is, therefore, significantly simpler to achieve the required scale of agitation by the mere change in the rotational speed of the impeller. This approach, in conjunction with geometric similarity attained with lab-scale and pilot-scale batches, is the most practical and scientific way of performing the scale-up operation. For instance, our group previously reported the scale-up of lipidic nanoparticles of quinapyramine sulfate by geometric similarity approach (Prayag et al., 2021).

Example A batch size of 10.0 mL of lab scale was increased in order to meet the scale-up of the process adopted to 200.0 mL batch size. Calculate the rotational speed (N) required to be kept in large-scale batch to obtain uniform and reproducible particle size as that of lab scale on the basis of geometric similarity principle.

Solution To compensate for scale-up effects and simulate lab-scale and pilot-size batches, rotational speed was determined using several approaches based on the geometric similarity approach and scale of agitation approach. There are several geometric parameters which need prior estimation before solving numerous equations about the scale of the agitation approach. For instance, the geometry of the vessel, the diameter of the impeller, and the height of liquid in the vessel were determined as shown in Fig. 10.2 and listed in Table 10.2.

After considering geometric similarity across scales, various parameters related to liquid motion were calculated by putting values in the following equations. Further the rotational speed (N) was calculated, and their values are presented in Table 10.3.

Table 10.2 Parameters of glasswares/instruments used across scales

Parameter	Lab-scale batch	Pilot-scale batch
Diameter of impeller (D) (cm)	2.06	4.15
Diameter of tank/ vessel (T) (cm)	3.10	7.83
Height of liquid in vessel (Z) (cm)	1.92	5.91
D/T	0.64	0.53
Z/T	0.61	0.75

Table 10.3 Calculation of rotational speed for lab- and pilot-scale batches using the scale of agitation approach

Parameter	Lab-scale batch	Pilot-scale batch
Terminal pumping number (N_Q)	0.443	0.561
Cross-sectional area of tank (A) (cm^2)	7.554	48.175
Effective pumping capacity (Q) (cm^3/sec)	45.163	288.423
Bulk fluid velocity (V_b) (cm/sec)	5.987	5.987
Impeller speed (rpm)	700	430

Terminal pumping number (N_Q)

$$N_Q = 1.1283 - 1.07118\left(\frac{D}{T}\right) \tag{10.18}$$

Cross-sectional area of tank (A)

$$A = \frac{\pi T^2}{4} \tag{10.19}$$

Effective pumping capacity (Q)

$$Q = N_Q \, x \, N \, x \, D^3 \tag{10.20}$$

Bulk fluid velocity (V_b)

$$V_b = \frac{Q}{A} \tag{10.21}$$

Table 10.3 lists various values calculated after application of geometric similarity principle in conjunction with scale of agitation approach across scales. The illustrative example provides insights into various processes involved in utilization of number of parameters together in achieving successful scale up process.

It can be observed from the table that the required impeller speed was drastically reduced from 700 to 430, in order to compensate scale-up effects and to maximize mixing efficiency across scales.

10.4.5 *Software-Based Approach of Scale-Up of Liquid Mixing Process*

Increasing the batch size or production scale of a mixing process is often necessary to meet production and clinical requirements, as well as to reduce production costs and time. Mixing liquids in a stirred tank can be a complex task, and it is crucial to understand the various parameters and their interactions to ensure the success of the mixing process. VisiMix® is a software-based simulator tool that can predict the hydrodynamics in the mixing vessel and understand various parameters such as power requirements, shear rates, and heat transfers. This can help design and optimize the mixing process, especially for large-scale preparations that involve mixing as a key phenomenon (Megawati et al., 2018). VisiMix® is a tool that takes into account several factors that are important for the successful scale-up of a mixing process. These factors include the calculation of the Reynolds number, the calculation of the P/V ratio, and the consideration of the principle of geometric similarity. By considering these factors, it is possible to design equipment that is suitable for scale-up, and that will produce consistent results at different scales. Megawati et al. have described the various calculations that are used in the scale-up process and have suggested that equalizing the P/V ratio across different scales may be a valuable approach for scaling up a mixing process (Megawati et al., 2018).

10.5 Conclusion

Liquid mixing is a critical unit operation in the manufacturing of various parenteral and non-parenteral liquid formulations. Still, there is a lack of information in the literature on the scale-up of liquid mixing processes. Among the various approaches available for scale-up, using the principle of similarity is a viable option. However, there is still a need for more research to improve the success of scale-up manufacturing liquid formulations. It is essential to understand the various factors that can affect the mixing process and to develop approaches and tools that can help to optimize the process and achieve consistent results at different scales.

References

Abadi, S. G. F. (2016). *The role of rheology in the flow and mixing of complex fluids*. Master of Philosophy Thesis. University of Birmingham.

Am Ende, D. J., & Am Ende, M. T. (2019). Chemical engineering in the pharmaceutical industry. *Chemical Engineering in the Pharmaceutical Industry*. https://doi.org/10.1002/9781119600800

Bathija, P. R. (2022). *Jet mixing*. Available at: https://www.mixing.com/jet-mixing. Accessed: 13 Oct 2022

Block, L. H. (2005). Scale up of liquid and semisolid manufacturing processes. *Pharmaceutical Technology*, s26–s33.

Bossler, F., et al. (2017). Influence of mixing conditions on the rheological properties and structure of capillary suspensions. *Colloids and Surfaces A: Physicochemical and Engineering Aspects, 518*, 85–97. https://doi.org/10.1016/J.COLSURFA.2017.01.026

Gates, L. E., & Henley, T. L. (1985). *Liquid Agitation*. Chemineer, Inc.

Hartmanshenn, C. (2020). Heat transfer in a bladed mixer: Scale up and effect of material and process parameters.

Heat Transfer and Temperature Difference of 2 liquids. (2013). Available at: https://physics.stackexchange.com/questions/106470/heat-transfer-and-temperature-difference-of-2-liquids

Kaalva, A. K. (2016). *Agitation scale up*. Available at: https://www.pharmacalculations.com/2016/05/agitation-scale-up.html

Kim, K., & Saito, S. (2013). Multiple length and time scales of dynamic heterogeneities in model glass-forming liquids: A systematic analysis of multi-point and multi-time correlations. *The Journal of Chemical Physics, 138*(12), 12A506. https://doi.org/10.1063/1.4769256

Kuksin, S., & Kuksin, S. (2021). *Kolmogorov's theory of turbulence and its rigorous 1d model* (Vol. 46, p. 181). HAL Open Science.

Lévêque, E. (2006). An introduction to turbulence in fluids, and modelling aspects. *EAS Publications Series, 21*, 7–42. https://doi.org/10.1051/eas:2006105

Lu, W. M., Wu, H. Z., & Ju, M. Y. (1997). Effects of baffle design on the liquid mixing in an aerated stirred tank with standard Rushton turbine impellers. *Chemical Engineering Science, 52*(21–22), 3843–3851. https://doi.org/10.1016/S0009-2509(97)88929-4

Luo, X., et al. (2021). Heat transfer and hydrodynamics in stirred tanks with liquid-solid flow studied by CFD–DEM method. *Processes, 9*(5). https://doi.org/10.3390/pr9050849

Megawati, et al. (2018). Scale-up of solid-liquid mixing based on constant power/volume and equal blend time using VisiMix simulation. *MATEC Web of Conferences, 187*, 1–4. https://doi.org/10.1051/matecconf/201818704002

Mixing and Mass Transfer | Reaction Rate – METLER TOLEDO. (2015). Available at: https://www.mt.com/in/en/home/applications/L1_AutoChem_Applications/L2_ProcessDevelopment/Mass-Transfer-and-Reaction-Rate.html. Accessed: 27 Oct 2022

Mixing: Mechanisms of Mixing. (2021). Available at: https://www.simplepharmanotes.com/2021/12/mixing-mechanisms-of-mixing.html. Accessed: 13 Oct 2022

Peters, S. (2015). *Introduction to mixer impellers & flow patterns*. Crane Engineering. Available at: https://blog.craneengineering.net/introduction-mixer-impellers-flow-patterns

PolyProcessing Solutions. (2022). *Using Baffles for Tank Mixing Applications*. Available at: https://blog.polyprocessing.com/blog/using-baffles-for-tank-mixing-applications. Accessed: 13 Oct 2022

Prayag, K. S., et al. (2021). Preparation and evaluation of quinapyramine sulphate-docusate sodium ionic complex loaded lipidic nanoparticles and its scale up using geometric similarity principle. *Journal of Pharmaceutical Sciences, 110*(5), 2241–2249.

Rushton, J. H. (2002). Mixing of liquids in chemical processing. *Industrial and Engineering Chemistry, 44*(12), 2931–2936. https://doi.org/10.1021/IE50516A045

The 6 Most Common Fluid Mixing Processes. (2022). Available at: https://www.linkedin.com/pulse/6-most-common-fluid-mixing-processes-mixquip/?trk=organization-update-content_share-article. Accessed: 13 Oct 2022

Torotwa, I., & Ji, C. (2018). A study of the mixing performance of different impeller designs in stirred vessels using computational fluid dynamics. *Designs, 2*(1), 1–16. https://doi.org/10.3390/designs2010010

Transvac. (2021). *Tank mixing with jet mixers – Transvac*. Available at: https://www.transvac.co.uk/tank-mixing/. Accessed: 13 Oct 2022

Part IV
Manufacturing and Scale-Up of Specialised Pharmaceutical Formulations

Chapter 11
Manufacturing Process of Nanoparticles

Meenakshi Kanwar Chauhan, Alisha Sachdeva, Lubna Ansari, and Dalapathi Gugulothu

Contents

Abstract Nanocarriers have emerged during the past few years as superior carriers and delivery systems for a variety of drugs with low bioavailability when compared to all other macromolecules. They also have exceptional traits like low toxicity, affordable scale-up manufacture, excellent biocompatibility, and large pharmaceutical loading efficacy. A variety of nanoparticle types are covered in this study, including dendrimers, solid lipid nanoparticles, polymeric nanocapsules, liposomes, etc. We have also discussed in depth these nanocarriers' composition, size, methods of synthesis, benefits, and some of their drawbacks. The most crucial drug delivery system strategy for the industrial formation of therapeutic compounds with poor water solubility is the nanosizing of pharmaceutical drug particles. Although drug particle nanosizing has been shown to significantly increase drug dissolving kinetics and apparent solubility, manufacturing of nanostructures as solid formulations has been significantly hampered (e.g. tablets, capsules). This is brought on by potent Van der Waals interactions among dry nanoparticles, which cause adhesion, agglomeration, and poor rheological properties. The vast field of nanomedicines is surveyed in this chapter, with a particular emphasis on top-down and bottom-up techniques.

M. K. Chauhan (✉) · A. Sachdeva · L. Ansari · D. Gugulothu (✉)
GOVT of NCT of Delhi, Delhi Institute of Pharmaceutical Sciences and Research (DIPSAR), Delhi Pharmaceutical Sciences and Research University (DPSRU), New Delhi, India
e-mail: dalapathig@dpsru.edu.in

A. B. Jindal (ed.), *Pharmaceutical Process Engineering and Scale-up Principles*, AAPS Introductions in the Pharmaceutical Sciences 13,
https://doi.org/10.1007/978-3-031-31380-6_11

Keywords Nanoparticles · Drug delivery systems · Production techniques · Bottom-up methods · Top-down methods

Abbreviations

BCS	Biopharmaceutical Classification System
CMC	Critical micelle concentration
MTX	Methotrexate
PLA	Poly (d, l-Lactide)
PVP	Polyvinyl Pyrrolidone
SARS CoV 2	Severe acute respiratory syndrome coronavirus
W/O/W	Water in oil in water

11.1 Introduction

The US FDA has authorized 100 nano-based drug formulations in the previous ten years (Etheridge et al., 2013), demonstrating the significance of nanotechnology in the present pharmaceutical sector. Nanoparticles are defined as molecules in the pharmaceutical sector that are smaller than 1 micrometre (Dilnawaz et al., 2018). Due to the severe drawbacks and issues with traditional active pharmaceutical components and completed dose formulations, nanotechnology has a huge effect on therapeutics and pharmaceutical delivery. 90% of the novel chemical substances in the pharmaceutical sector's drug discovery pipelines (Tan et al., 2017). Because there is little accessible diffusion due to the drug's limited solubility at the location of administration, there is inadequate API amount at the target, which results in in vivo failure (Siepmann & Siepmann, 2013). Class I (high solubility and high permeability), Class II (low solubility and high permeability), Class III (high solubility and low permeability), and Class IV (low solubility and low permeability) are the four types that the Biopharmaceutical Classification System (BCS) differentiates pharmaceutical drugs based on their aqueous solubility and intestinal permeability (Amidon et al., 1995). The BCS approach places a lot of emphasis on a drug's aqueous solubility since it's crucial for absorbing drug moieties that are passively carried over the gastrointestinal tract (Devalapally et al., 2007). A drug is deemed to be very soluble, according FDA recommendations, if the maximum dose strength of the drug can be dissolved in 250 mL of aqueous medium with pH ranging from 1 to 6.8, at a temperature of 37° ± 1 °C. Whereas a drug is said to be extremely permeable if greater than 85% of it is absorbed in the body after administration (Rautio et al., 2008; Charalabidis et al., 2019).

Till now, the issue of less solubility has been resolved by adopting micronization procedures for particle size reduction of drugs with low solubility (Aguiar et al., 2017). However, a switch from micronization to nanonization has occurred due to the necessity for additional advancements in medication dispersion. Much interest has grown recently in creating colloidal nanoparticles with altered biological characteristics that allow for altered drug delivery and targeting. The organic nanoparticles may be classified as liposomal and lipid-based, polymeric and core-shell, micellar, dendrimer, and drug nanocrystals based on their content, function, and appearance (9). As they are biocompatible, liposomes potentially entrap both hydrophobic and hydrophilic pharmaceuticals in their membrane and internal water division, respectively, therefore shielding the medicines from the deactivating effects of environmental factors without resulting in harmful side effects (Torchilin, 2005). Because of their capacity to improve the solubility and therapeutic index of chemotherapeutic drugs, liposomes are intensively researched in the drug delivery sector (Torchilin, 2005). Amphotericin B and doxorubicin liposomal formulations served as the foundation for the first nanomedicines that the FDA cleared for clinical studies in the middle of the 1990s (Vaage et al., 1992).

Various techniques, including nanoprecipitation (Zili et al., 2005), double emulsification (Jeong et al., 2008), emulsion-diffusion (Guinebretière et al., 2002), and emulsion coacervation (Lertsutthiwong et al., 2009), have been used for the development of polymeric nanoparticles. Typically, polymeric nanomedicines fall into one of two groupings: either controlled release applications with degradable polymer structure or polymer-drug conjugates for improved half-life and bioavailability (Bobo et al., 2016). Non-ionic surfactants are used to coat these nanocapsules, which lessens immunological interactions such opsonization. Copaxone® is the greatest example of a polymeric nanoparticle medication that has received FDA approval as well as one that is now through a clinical study. Copaxone®, which functions as an immunomodulatory for multiple sclerosis, received its first approval in 1996 (Johnson et al., 1998).

Polymeric micellar systems are self-assembled polymeric amphiphiles of ABA triblock or AB diblock copolymers designed for the controlled release of hydrophobic medicines in an aquatic environment. In comparison to surfactant-based micelles, the usage of block copolymers results in a lower critical micelle concentration (CMC) and greater stability (Oerlemans et al., 2010). Although the polar hydrophobic external surface of the micelles facilitates dissolution in an aqueous media, the hydrophobic core of the micelles is employed close to enclose poorly soluble drugs, therefore reducing their removal from the body (Hammad & Müller, 1998). The micelle's core-shell structure replicates the body's natural transport mechanism, making it easier for drugs to be absorbed and distributed (Kwon & Okano, 1996; Kabanov & Alakhov, 2002). The only micellar formulation on the market that has received FDA approval to treat moderate to severe vasomotor symptoms of menopause is estradiol (Estrasorb™) (Bobo et al., 2016). Additionally, there are several clinical trials for micellar formulations. For instance, the treatments for prostate cancer have significantly improved because of BIND-104. It is a micellar formulation with docetaxel enclosed in a hydrophilic PEG shell and a hydrophobic polymeric core that may be broken down (Hrkach et al., 2012). The micellar formulations CriPec® (Rijcken et al., 2005) and CALAA-01 are two further examples that are presently undergoing clinical studies (Davis, 2009).

Dendrimers are highly branched, three-dimensional nanoparticles that end in a number of surface functional groups (Tomalia et al., 1990). Dendrimers are utilized to boost the solubility of a hydrophobic medication either by encasing them in their intramolecular cavity or by connecting to their external surface functional groups, due to their greater solubility (Soto-Castro et al., 2012) and bioavailability (Duncan & Izzo, 2005). The commercially available dendrimers or dendrimer-based products include Priostar® (Swanson et al., 2007), Starburst® (Tomalia et al., 1990), and Polylysine (Patton et al., 2006). Due to their distinctive superparamagnetic and magnetic responsiveness as well as their characteristics of nanoparticles, such as a nano-level particle size and huge specific surface area, magnetic nanomaterials have received the most attention among these nanoparticles. In a continuous magnetic field, magnetic nanoparticles can congregate in a particular area. In order to produce heat, it can also vibrate rapidly in an alternating magnetic field. Due to their benefits, they are a very promising and useful research nanoparticle for targeted cancer therapy (Bai, 2022). Superparamagnetic iron oxide nanoparticles, often known as MNPs, have been extensively employed as contrast agents in magnetic resonance imaging throughout the past few decades. The use of MNPs in medicine has increased dramatically in recent years (Yallapu et al., 2010). A cutting-edge method for precisely delivering medications to tumour tissue is magnetic targeting (Veiseh et al., 2010). When combined with MNPs on the surface of tumours, magnetic force can lengthen the drug's surface retention duration. MNPs may be divided into three categories: monometallic, bimetallic, and magnetic alloy nanoparticles (Zhang et al., 2013). Iron nanoparticles like Fe_3O_4 and Fe_2O_3 are monometallic nanoparticles. Among the bimetal particles are mostly $CoFe_2O_4$ and $ZnFe_2O_4$ (Li et al., 2021).

One of the primary uses of MNPs in the treatment of cancer is targeted drug delivery. Some tumours, like pancreatic cancer, are lodged in the organs because the site of the disease is unknown, which makes it challenging for drugs that ordinarily circulate in the blood to reach the lesions. Broad-spectrum anticancer medicines including doxorubicin, methotrexate, cisplatin, and docetaxel make up the majority of the anticancer medications that have been utilized in clinical practise (Schmid et al., 2006). They frequently lack targeting and specificity and are simple to circulate constantly in the body. They can unnecessarily harm healthy tissues when delivering the drug. MNPs are an efficient substitute for targeted medication delivery because they can get over these biological obstacles and restrictions (Bai, 2022).

Another approach for drug delivery based on nanotechnology is drug nanocrystals. They stand out because they are 100% drug and have higher solubility owing to expanded surface area, which increases saturation solubility and reduces mucosal layer retention. Additionally, nanocrystals boost dose bioavailability, oral absorption reproducibility, and patient compliance by reducing the number of administrations (Müller et al., 2001). Drug nanocrystals can be administered intravenously as aqueous nanosuspensions in addition to being a spectacular oral drug delivery technology (Hanafy et al., 2007). A nanocrystal formulation typically consists of drug nanoparticles distributed in an aqueous or non-aqueous solution together with one or more stabilizers. The very first lipid-based pharmaceutical

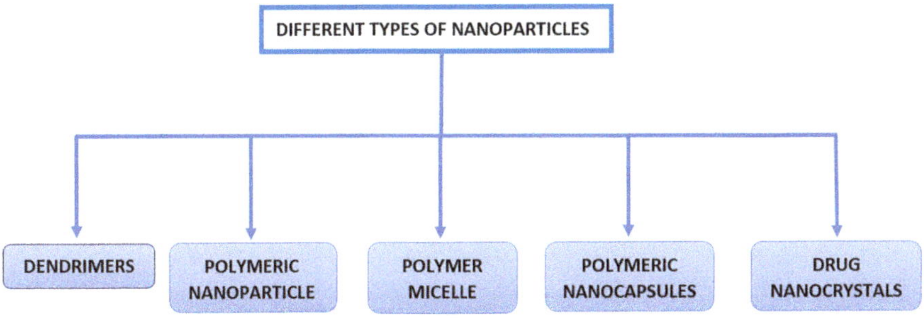

Flowchart 11.1 Depicting various types of nanoparticles

preparation took around 25 years to reach the market, but the nanocrystal-based product Emend® did so within 10 years after the first patent filing. Drug nanoparticles can be created using bottom-up or top-down methods (diminution strategy or precipitation method). The different types of nanoparticles are shown in Flowchart 11.1 and Fig. 11.1 (Khodabandehloo et al., 2016).

Although nanoparticles have many benefits, several disadvantages include stability (Patravale et al., 2004; Rabinow, 2004), difficult production (Blagden et al., 2007), and nanotoxicity (Fischer & Chan, 2007). The most crucial factor in guaranteeing the efficiency and safety of pharmacological products is stability. Physical stability and chemical stability are two kinds of stability. Physical stability problems include sedimentation, agglomeration, crystal formation, and recrystallization. Chemical stability problems arise from functional group decomposition, like the amides and esters by hydrolytic breakdown and amino groups by the oxidative decomposition. Chemical stabilities are specific to drug, in contrast to the physical stability problems that are frequently seen. The development of bigger particles (>5 m) from intravenously delivered nanosuspensions is an instance of agglomeration or crystal growth that may occur. These particles may cause capillary obstruction and embolism (Patravale et al., 2004).

Manufacturing, storage, and transportation processes for pharmaceutical products are impacted by physical stability. The rheological characteristics of formulations are also impacted by these problems with physical stability. Due to their charged nature, nanocrystals can have led flow ability and compressibility, which causes their aggregation or build crystals. As a result, the therapeutic formulation is ineffective, which ultimately results in failure of treatment throughout the clinical phases. Therefore, there exists a huge need in the pharmaceutical industry to separate drug nanoparticles by evenly encapsulating them onto other micron-sized carriers, or encasing them in polymer matrices, therefore optimizing rheology in therapeutic formulation (Verma et al., 2021).

Manufacturing of Nanoparticles

Nanoparticle can be manufactured by bottom-up method, top-down method, and scale-up method and each method was discussed in detail.

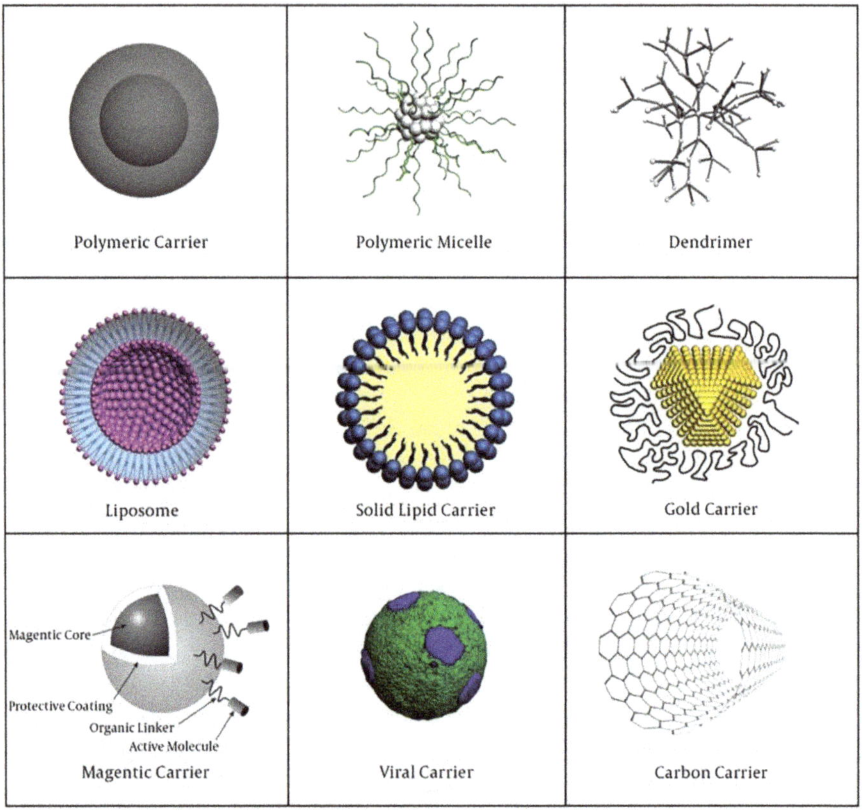

Fig. 11.1 Diagrammatic representation of various kinds of nanoparticles. (Khodabandehloo et al., 2016). (This is an open-access article distributed under the terms of the Creative Commons Attribution-NonCommercial 4.0 International License (http://creativecommons.org/licenses/by-nc/4.0/) which permits to copy and redistribute the material just in noncommercial usages, provided the original work is properly cited) (Flowchart 11.2)

11.2 Bottom-up Methods

Most frequently, in bottom-up methods, pharmaceutical nanoparticles are precipitated in crystalline or amorphous form from a supersaturated solution which can be aqueous or organic, either by the solvent evaporating or by the application of an anti-solvent (Sinha et al., 2013; Salazar et al., 2014).

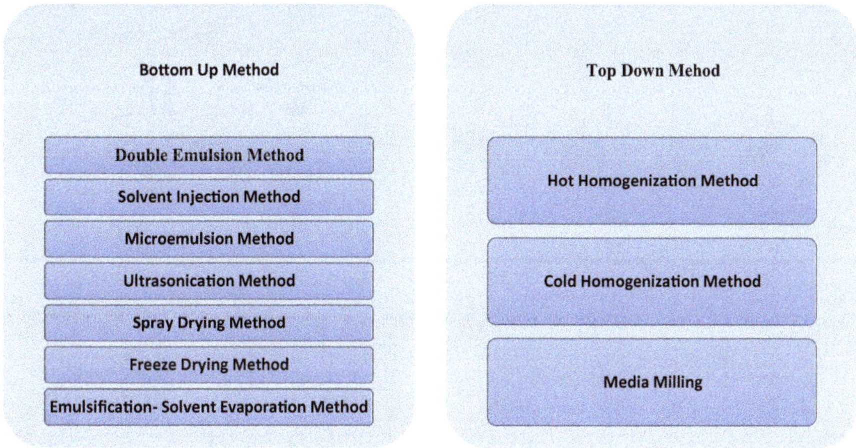

Flowchart 11.2 Various techniques of production of nanoparticles via bottom-up and top-down method

11.2.1 Double Emulsion Method

Double emulsion approach works well for water-soluble active pharmaceutical ingredients. Water-soluble pharmaceuticals should be first dissolved in an aqueous solvent, followed by combining the solution with the melted solution of lipid to form a single solution. A double emulsion method is an efficient way to encapsulate hydrophilic drugs in lipid carriers. Primarily, after dissolving said drugs in an aqueous solvent, and emulsion surfactants containing a secondary aqueous solution stabilizes the formed single emulsion. It is then combined by hydrophilic stabilizer aqueous solution to form an emulsion in two parts. As a result, the formed double emulsion is purified in cold conditions by centrifugation or filtration techniques. Because this approach employs water-soluble surfactant, the alteration of surface is simple to carry out. The obtained particle size in this method is greater and necessitates additional treatment (Svilenov & Tzachev, 2014). Polyethylene glycol nanoparticles have been developed using double emulsion technique. Also, cetyl palmitate solid lipid nanoparticles (SLNs) were tried using this method (McKay et al., 2020). Using an ultrasonic probe, a customized water-in-oil-in-water (W/O/W) emulsification and solvent evaporation procedure was used to create poly(d, l-Lactide) (PLA) nanocapsules (Zhu et al., 2005). On the other hand, nanoparticle-in-microparticle (NIM) was synthesized using the double emulsification method, which is a combination of single component particles such as nanoparticles and microparticles. Their combination allows for dual or multiple functionalities, such as multiple release profiles, within a formulation (Lee et al., 2013).

The fabrication of a series of DNR-loaded nanoparticles in poly(D, L-lactic-co-glycolic acid (PLGA) and poly(D, L-lactic acid) (PDLLA) was done utilizing a modified double-emulsion solvent evaporation/diffusion process that included a

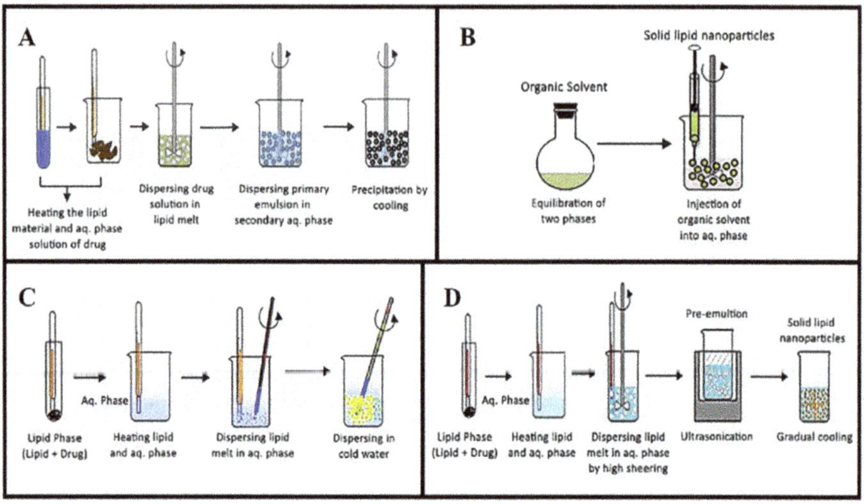

Fig. 11.2 Various methods for synthesizing nano-lipid-based carriers are shown schematically: (**a**) Double Emulsion Method (**b**) Solvent Injection Method (**c**) Microemulsion Method (**d**) Ultrasonication Method. (Ganesan & Narayanasamy, 2017) (Copyright 2022, Elsevier)

partially water-soluble organic solvent. In comparison to free DNR, optimized DNR/PLGA nanoparticles demonstrated significantly increased cellular absorption and greater cytotoxicity against HL-60 cells. This approach may be helpful for hydrophilic chemotherapeutic medicines that need to be delivered to cancer cells effectively and then released over time at a specific site (Liu et al., 2010).

Another anticancer agent is Nisin, and it is a peptide that has antibacterial and anticancer effects, but one of its major drawbacks is that it degrades quickly by enzymatic means and permeates the cell membrane only to a limited extent. In order to effectively cure cancer without having a substantial negative impact on healthy cells, NPN was created using a double emulsion solvent evaporation process (Haider et al., 2022). This method is represented in Fig. 11.2 (Guimarães et al., 2020).

11.2.2 Solvent Injection Method

This method uses simple methods to produce lipid nanoparticles and is based on the diffusion principle. First, solvents like methanol, acetone, and isopropanol are used for dissolving lipid to get the organic phase then injected into a surfactant-containing aqueous solution using a syringe needle which is diagrammatically represented in Fig. 11.2 (Guimarães et al., 2020).

The following two factors have an impact on lipid nanoparticles synthesis.

(a) The progressive removal of solvent from lipid droplets decreases droplet size while enhancing lipid concentration.

(b) Once the solvent is evaporated from the lipid droplets, the interfacial tension at the surface of the droplets enhances, hence droplets contract still more.

Changes in process parameters can also influence particle size (Schubert & Müller-Goymann, 2003).

Recently, a new Methotrexate (MTX)-loaded nanocarrier to treat rheumatoid arthritis utilizing the ethanol injection technique was developed employing the liposomal formulation, comprising of a 1:1 ratio of organic: aqueous phase (v/v), and the phospholipids DOPE/cholesterol/DSPE-mPEG2000 (Guimarães et al., 2020). Liposomes comprising of phosphatidylcholine and 1,2-dipalmitoyl Sn-glycero-phosphocholine monohydrate were also developed (Talluri et al., 2016). Using the solvent injection approach, the water-soluble medication ondansetron hydrochloride was added to nanostructured lipid carriers to enhance the drug's pharmacokinetics. Using the right amount of liquid lipid produced tiny, monodispersed NLCs with improved EE (entrapment efficiency) and drug loading. The improved NLCs formulation displayed an in vitro sustained-release profile of ODS along with particle size of 185.2 ± 1.9 nm, polydispersity index of 0.214 ± 0.006, EE of 93.2 ± 0.5%, and DL of 10.43 ± 0.05% (Duong et al., 2019). Similarly, doxorubicin-hydrochloride nanoliposome (DHNP) preparation method using ethanol injection and a pH gradient has the benefit of uniform distribution of size, greater encapsulation efficiency, high drug loading rate, and reduced toxicity than free doxorubicin (Xie et al., 2013).

Paclitaxel-cholesterol complex-loaded lecithin-chitosan nanoparticles (PTX-CH-loaded LCS NPs) were made using a solvent-injection technique for intratumoral injection treatment. A palliative treatment known as intratumoral injection tries to further improve survival of cancer patients having advanced or recurring carcinomas, significant comorbidities, or low performance status. It presented an alternate strategy in the realm of palliative chemotherapy by demonstrating enhanced safety and antitumor activity (Chu et al., 2019).

11.2.3 Micro Emulsion Method

Because of the low energy requirement, this method is used to produce lipid nanoparticles. It is being used concurrently; the lipid is heated and subsequently mixed with a pre-heated aqueous phase with gently stirring to form a microemulsion. It is then poured into a large volume of cold water in the second step to solidify the droplets under reasonable stirring conditions as shown in Fig. 11.2 (Guimarães et al., 2020). The obtained nanoparticles are spherical in shape and small in size. This process needs a highly diluted solution due to which it requires evapoartion at later stage to obtain particles in concentrated form (Svilenov & Tzachev, 2014). Nano-structured lipid carriers comprising of oleic acid and squalene nanoparticles have been created utilizing this approach (Fang et al., 2013). To make calcium phosphate CaP/pDNA nanoparticles, a Triton X-100/Butanol/Cyclohexane/Water reverse microemulsion technique was employed. The preparation of microemulsion A involved the use of TritonX-100 and Butanol as mixed emulsifiers, hexane as the oil phase, 100 microlitre of calcium chloride solution (1.0 M), 500 ng pDNA, and 1 h of stirring. In a similar manner, microemulsion B was created by adding 100 l of Na_3PO_4 (1.0 M) to the oil phase and stirring it for 1 h. Microemulsion A was combined with microemulsion B, and the mixture was then agitated for an additional 12 h at a specific speed. The created NPs are ph sensitive, have good biocompatibility, and have good encapsulation efficiency (Li et al., 2017). The aqueous cores of sodium bis(2-ethylhexyl) sulfosuccinate (AOT)/n-hexane reverse micelles were used to create glycidyl methacrylate derivatized dextran (Dex-GMA) nanoparticles that contained basic fibroblast growth factor (bFGF). Dex-GMA nanoparticles were made by combining water-in-oil micro-emulsion systems with an aprotic solvent, and they were found to be consistently spherical in shape with an average size of 109.57+/−2.09 nm. They represent a new, efficient, and biocompatible delivery vehicle for bioactive protein (Wu et al., 2009).

11.2.4 Ultrasonication

This technique uses dispersion technique which makes use of great energy to form lipid-based nanoparticles. The melted lipid solution is added to the pre-heated aqueous solution with a magnetic stirrer for the formulation of the emulsion. Following that, large droplets of emulsion are formed. These are further reduced using ultrasonication at a specific amplitude utilizing containers providing ice conditions. The emulsion is then slowly cooled, which led to the development of lipid nanoparticles. Following that, ultracentrifugation was used to purify the sample (Shah et al., 2015) as depicted in Fig. 11.2 (Guimarães et al., 2020). Glycerol tristearate and monostearate SLNs have been formulated by this method (McKay et al., 2020). Insoluble rice peptide aggregates that are deemed undesirable and develop during the ultrasonic hydrolysis of rice protein have been converted into a novel food-grade pickering stabilizer. Additionally, Fe^{2+} chelating activity is promoted by ultrasonic treatment, which can enhance rice peptide nanoparticle (RPNs) antioxidant activity. These results offer both a unique bifunctional pickering stabilizer with inherent antioxidant capabilities and a fresh strategy for the efficient exploitation of insoluble aggregates generated during protein hydrolysis (Zhang et al., 2021). Another example is the production of starch nanoparticles from Short linear glucan debranched from waxy maize starch and SNPs that were peppermint oil (PO)-loaded. The as-prepared SNPs had diameters of 150–200 nm, high homogeneity, and a nearly flawless spherical shape. The yield, encapsulation effectiveness, and loading capacity of PO-loaded SNPs were, respectively, 25.5%, 87.7%, and 93.2%. The slow-release characteristics of PO from SNPs were precisely predicted by the pseudo-first-order kinetics model. This new method of making SNPs is quick, high yielding, and non-toxic, and it has a lot of promise for encapsulating and sustaining the release of lipids like volatile essential oils (Liu et al., 2017).

11.2.5 Freeze-Drying Method

Freeze-drying, also known as lyophilization, is a manufacturing procedure that uses a vacuum system to eliminate water from a frozen sample by sublimation and desorption. Using this drying process, labile ingredient solutions are transformed into solids that seem to be stable enough to be transported and stored. The most common technique for extending the shelf life of thermosensitive medications included within NLBCs is freeze-drying. Sucrose, hydroxypropylcyclodextrin, and trehalose are some cryoprotectants that further improve the nanocarriers stability while processing this cycle. As a result, stability plays a crucial role in guaranteeing the efficacy and safety of medications (Kumar et al., 2022). Polymorphs of piroxicam were converted into nanocrystals by Lai et al. using the HPH technique, and Poloxamer 188 was used to stabilize the nanosuspension. Xanthan gum, PEG 400, and maltodextrins were used as cryoprotectants during the freeze-drying process to separate the nanocrystals from the nanosuspension. Because of the high energy involved in the HPH process, X-ray diffraction demonstrated the polymorphic change of piroxicam from form I to form III and monohydrate. Piroxicam nanocrystals had a higher surface-to-volume ratio than coarse piroxicam, which boosted the rate of dissolution (Lai et al., 2011). By employing the freeze-drying method with sucrose as the matrix forming, Liversidge et al. patented the solidification of polyvinylpyrrolidone (PVP) stabilized danazol nanosuspension. The danazol particle has an approximate nanocrystal size of 400 nm (Verma et al., 2021). Additionally, sucrose was employed as a cryoprotectant and matrix forming for the solidification of loviride nanosuspension, which was made utilizing a media milling method using zirconium beads with a 0.5 mm diameter as the milling agent (Van Eerdenbrugh et al., 2007). Freeze-drying approach of manufacturing of nanoparticles suspension and obtaining dry powder from it is depicted in Fig. 11.3 (Chu et al., 2019).

11.2.6 Spray-Drying Method

To form an emulsion, the lipid is mixed in the same solvent as the drug. Following that, the solution is dried by passing it through the spray-drying chamber's atomizer by using hot drying gas. Nitrogen gas is used in this system because the organic solvent is used to keep the environment inert used during the method. Furthermore, the liquid spray is used in the drying process compartment with hydraulic, ultrasonic, rotary, and other nozzles and fluid nozzles. Whenever the spray dispersion comes into contact with a hot drying gas, the solvent begins to evaporate, causing the sample to dry. The particles then evaporated from the gas stream using a cyclone separator. Though many industries use spray dryers in recycling mode to produce lipid nanoparticles on a huge scale (Dobry et al., 2009). This technique was used to create spray-dried LNPs for the delivery of RNA medications via dry powder

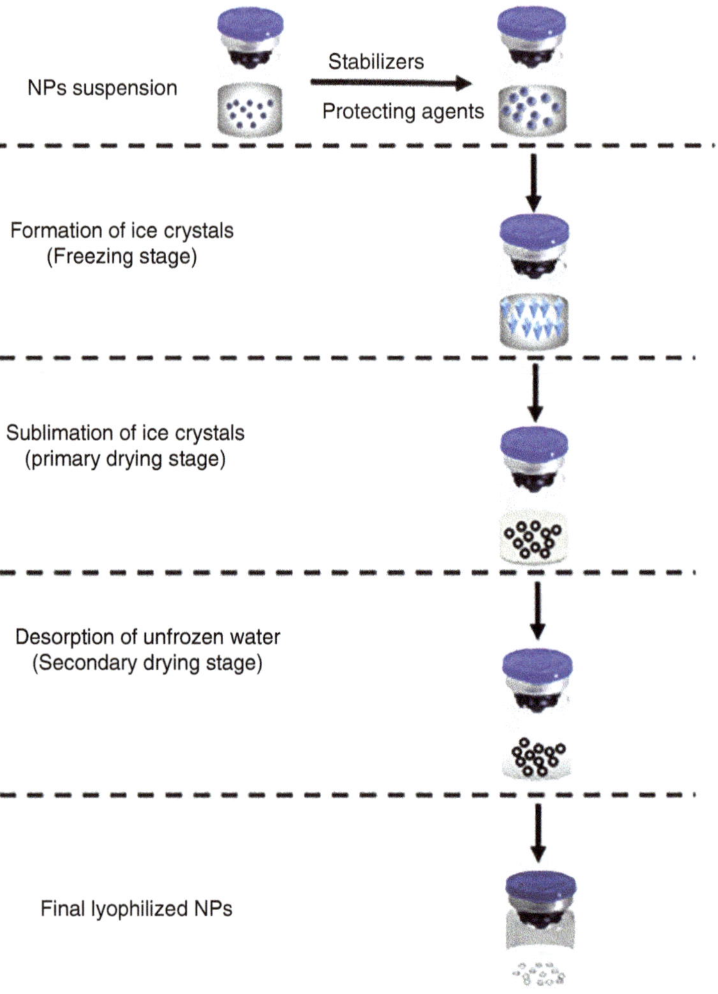

Fig. 11.3 The varying stages in freeze-drying of nanoparticles and the phase transitions of product water through different stages of the process. (Mohammady et al., 2020) (Copyright, 2020, Elsevier)

inhalers, which has significant benefits for the physical, chemical, and microbiological stability of RNA and nanosuspensions. The lungs provide a variety of now unreachable targets that could possibly be treated with RNA therapies, whereas all siRNA medications currently on the market target the liver. In a lung cancer cell line, spray-dried LNPs downregulated protein by >90% and pierced the lung mucus layer while maintaining bioactivity (Zimmermann et al., 2022). Similarly, Vildagliptin (445 nm) and calpain inhibitor (300 nm) have been developed successfully by nanospray technique (Harsha et al., 2015). This method is pictorially depicted in Fig. 11.4a (Guimarães et al., 2020)

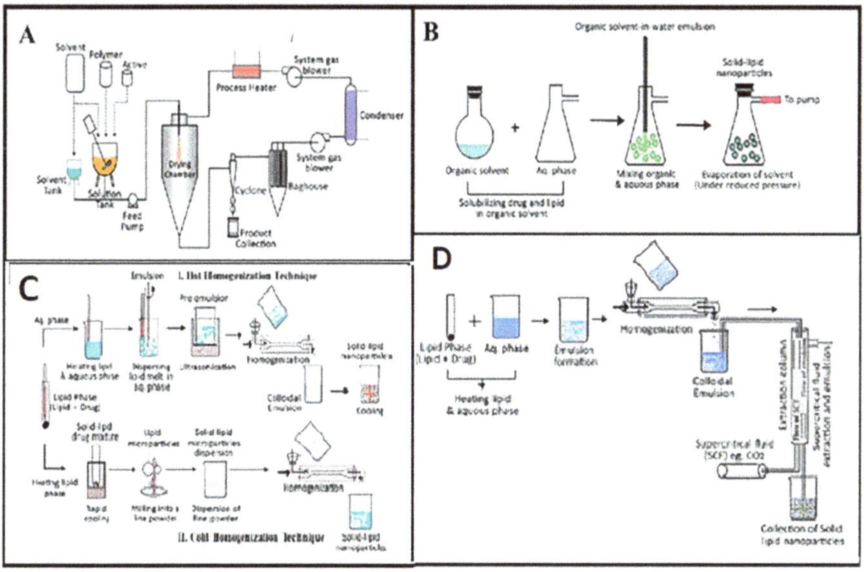

Fig. 11.4 Various methods for the manufacture of nano-lipid-based carriers are depicted schematically. Methods of (**a**) Spray drying and (**b**) Emulsification and solvent evaporation (**c**) Hot/Cold homogenization (**d**) Method of supercritical fluid. (Ganesan & Narayanasamy, 2017) (Copyright 2022, Elsevier)

11.2.7 Emulsification and Solvent Evaporation

This method constitutes three phases for producing lipid nanoparticles:

1. Organic phase preparation.
2. Pre-emulsification.
3. Nano-emulsification (Pedersen et al., 2006).

First, lipids and hydrophobic drugs are dissolved in an organic phase, which is then added to the aqueous phase using a high-speed homogenizer to form particle droplets. Then the particle's formed droplets are reduced to the nanometre range using a high-pressure sonicator/homogenizer. Afterwards, the nanoparticle dispersion is kept on a magnetic stirrer immediately to vanish the solvent as shown in Fig. 11.4b (Guimarães et al., 2020). The method illustrated above can be used to develop nanoparticles by enhanced physicochemical properties (Jaiswal et al., 2016). SLNs have been designed using this method. Lipid polymer hybrid nanoparticles were created using lipids like Lecithin/DSPE/DLPC/DMPE/Cholesterol (Paliwal et al., 2010; Pardi et al., 2018).

For the prevention and control of illnesses and pests, formulations of safe and effective pesticides have received significant attention. A focus of current research in the field of pesticide formulations has been on using nanotechnology to extend the potency and longevity of insecticides. For example, ultrasonic emulsification-solvent evaporation method was used to create novel hydrophilic lambda-cyhalothrin nanospheres enclosed in poly(styrene-co-maleic anhydride; PSMA), which demonstrated superior particle size homogeneity and dispersion than the conventional method. This study opens up new possibilities for plant protection and residue reduction by developing a unique nano-formulation with hydrophilicity and high leaf adherence (Wang et al., 2022).

11.3 Top-Down Methods

Top-down strategies utilize high-energy methods to diminish medication particle sizes all the way down to the nanoscale. Media milling and high-pressure homogenization are the two major top-down method subgroups. Both of these methods used for reducing the size of the particles rely on collisions, tension, attrition, and shearing forces (Möschwitzer & Müller, 2006; Gao et al., 2013; Al-Kassas et al., 2017).

11.3.1 Hot Homogenization Method

Lipids and drugs are dissolved in an organic solvent and then mixed with an aqueous surfactant solution. A pre-emulsion formation tends to happen when melted lipid is dispersed in an aqueous phase utilizing a shear device. Following that, the pre-emulsion was ultrasonicated, and the hot homogenization is applied. Then homogenized mixture was allowed to cool at room temperature in order to get the droplets of liquid crystals (Svilenov & Tzachev, 2014) as shown in Fig. 11.4c (Guimarães et al., 2020). Triglycerides nanostructured lipid particles of all-trans retinoic acid were formulated using hot homogenization method (Fang et al., 2013).

Combinational preparations are used to create the valacyclovir SLNs. The nanoparticles were created using a high-pressure hot-homogenization method. Human viral infections, especially those brought on by members of the herpes virus family, are frequently treated and prevented with the use of the drug valacyclovir. A good medication for treating cold sores is valacyclovir. With antioxidant, hydrating, and stabilizing properties, jojoba *Simmondsia Chinensis* oil is one of the primary constituents. Additionally, grape seed (*Vitis vinifera*) oil, which has excellent therapeutic qualities including antioxidant, astringent, and moisturizing characteristics, is also utilized in this mixture (Chacko & Newton, 2019).

11.3.2 Cold Homogenization Method

It is used to address issues associated with the hot homogenization process, like drug degradation and nanoemulsion crystallization problems (Mehnert & Mäder, 2001). This method is similar to 2.1.1. However, in this process liquid nitrogen rapid cooling is employed to solidify the molten lipids and drugs (Jaiswal et al., 2016). Drugs are evenly spread within carriers made of lipids during the fast cooling process. The lipid solid particles are then converted into fine particles by the process of milling, which are next dispersed in an aqueous surfactant-containing solution, and finally, the homogenizer is used to produce the lipid nanoparticles from the dispersed solution. Vitamin E, stearic acid, tristearin, and cholesterol are some of the compounds whose nanoparticles are formulated using this approach (Fang et al., 2013).

Myricitrin, a plant-derived antioxidant, was used to create SLN, which could be more effective for treating Type 2 diabetes mellitus (T2DM), Oxidative stress could be a factor in T2DM. Myricitrin is an antioxidant that comes from plants, and its SLN may be more effective. In vivo and in vitro research using myricitrin SLN show improved consequences related to diabetes and hyperglycaemia (Ahangarpour et al., 2018).

Preparation of hydrophilic streptomycin sulphate (STRS; log P -6.4) in a lipid matrix of SLNs at a high dose (1 g/day) by the cold high-pressure homogenization technique used for SLN preparation achieved with 30% drug loading and entrapment efficiency of 51.17 0.95%. Polyethylene glycol 600 was assigned a small size (218.1 15.46 nm) with mucus-penetrating properties used as a supporting surfactant. The microassay demonstrated significant oral absorption and bioavailability higher than the free drug (Singh et al., 2021).

A cold homogenization technique was used to create SLNs loaded with vinorelbine bitartrate. The average particle size of the SLNs ranged from 150 to 350 nm. The drug release demonstrated that the drug release could last for 48 h, and the rate was reduced by lecithin or oleic acid addition to the formulations (You et al., 2007). This process is shown in Fig. 11.4d (Guimarães et al., 2020).

11.3.3 Media Milling

Media milling, which can be both dry and wet media milling, is also known as mechanochemical process (Rasenack & Müller, 2004; Merisko-Liversidge & Liversidge, 2011). Liversidge et al. invented this technique in 1992, and it is still in use today commonly referred to as Nanocrystal® (Möschwitzer, 2013). Dry media milling diminishes size of substances by utilizing high-energy interactions between particle-particle or particle-wall collision caused by an opposing high-velocity jet air that has been compressed (Louey et al., 2004). Wet milling, on the other hand, utilizes low energy approach and in this technique the drug is dispersed in the aqueous or non-aqueous medium, as well as a trace of stabilizer, followed by milling with hard glass, ceramic, or stainless steel beads (Leleux & Williams, 2014; Li et al., 2016). Milling approaches have been highly investigated in the literature to decrease the size of the particle of API to improve the pharmacokinetic attributes. Using a nanocrystal solid dispersion, Onoue et al., developed tranilast, a new anti-allergic drug that has been employed in the asthma therapy. The wet milling method was used to develop tranilast nanocrystals with a mean particle size of 122 nm. For the management of asthma, these nanocrystals offered superior dissolving behaviour and a reduced systemic concentration as a substitute to oral treatments (Onoue et al., 2011).

11.4 Conclusion and Future Perspective

In this chapter we have discussed various methods of nanoparticles manufacturing process for pharmaceutical drugs. The need of studying nanotechnology arises because of the low solubility of many drugs which has become a main difficulty (poor bioavailability and absorption) in the drug development process. As drug particle size decreases by nanosizing, surface area increases, which proportionally increases dissolution rate and saturation solubility. Nanoparticles loaded with drugs target specific sites that not only avoid first-pass metabolism but also decrease the drug toxicity level. Recently, there has been an increase in demand for anti-microbial or anti-bacterial materials due to the discovery of contagious viruses and bacteria. They spread exponentially through mutations and have developed immunity to many antibiotics and other drugs. Hence, there is a need for an alternative mechanism to deactivate their severity and thus reduce their effects. Silver nanoparticles have a lot of potential in terms of anti-microbial properties by deactivating the viruses and thus preventing their further mutation.

There are two main approaches for the preparation of drug nanoparticles, the top-down and bottom-up techniques. The growing demand from the pharmaceutical industry is driving the growth of the global NP market. Though many nanoscale processes are available to the pharmaceutical industry, only a few formulations comprising nanocrystals of API can reach the clinical trial stages. The possible reason for this could be (1) nanosizing principle is a complex process, (2) lack of enough in-house abilities related to technology and expertise for clinical trials, (3) to poor rheological properties of nanoformulations, and (4) lack of formal regulation for production of nanomaterials. It could be dealt by collaborative work and effective communication among researchers with different backgrounds, globalization of rules and regulations on nanomaterial production, and development of advanced preparation techniques such as microfluidics, electro spraying, etc. is needed to improve productivity and controllability.

References

Aguiar, G. P. S., et al. (2017). Micronization of N-acetylcysteine by supercritical fluid: Evaluation of *in vitro* and *in vivo* biological activity. *The Journal of Supercritical Fluids, 130*, 282–291. https://doi.org/10.1016/j.supflu.2017.06.010

Ahangarpour, A., et al. (2018). Solid lipid nanoparticles of myricitrin have antioxidant and antidiabetic effects on streptozotocin-nicotinamide-induced diabetic model and Myotube cell of male mouse. *Oxidative Medicine and Cellular Longevity, 2018*. https://doi.org/10.1155/2018/7496936

Al-Kassas, R., Bansal, M., & Shaw, J. (2017). Nanosizing techniques for improving bioavailability of drugs. *Journal of Controlled Release, 260*, 202–212. https://doi.org/10.1016/j.jconrel.2017.06.003

Amidon, G. L., et al. (1995). A theoretical basis for a biopharmaceutic drug classification: The correlation of in vitro drug product dissolution and in vivo bioavailability. *Pharmaceutical Research, 12*(3), 413–420. https://doi.org/10.1023/a:1016212804288

Bai, Y. (2022). *The application of magnetic nanoparticles as a targeted therapy in cancer treatment BT – Proceedings of the 2022 international conference on social sciences and humanities and arts (SSHA 2022)* (pp. 194–197). Atlantis Press. https://doi.org/10.2991/assehr.k.220401.039

Blagden, N., et al. (2007). Crystal engineering of active pharmaceutical ingredients to improve solubility and dissolution rates. *Advanced Drug Delivery Reviews, 59*(7), 617–630. https://doi.org/10.1016/j.addr.2007.05.011

Bobo, D., et al. (2016). Nanoparticle-based medicines: A review of FDA-approved materials and clinical trials to Date. *Pharmaceutical Research, 33*(10), 2373–2387. https://doi.org/10.1007/s11095-016-1958-5

Chacko, A., & Newton, A. M. J. (2019). Synthesis and characterization of valacyclovir HCl hybrid solid lipid nanoparticles by using natural oils. *Recent Patents on Drug Delivery & Formulation, 13*(1), 46–61. https://doi.org/10.2174/1872211313666190304142129

Charalabidis, A., et al. (2019). The biopharmaceutics classification system (BCS) and the biopharmaceutics drug disposition classification system (BDDCS): Beyond guidelines. *International Journal of Pharmaceutics, 566*, 264–281. https://doi.org/10.1016/j.ijpharm.2019.05.041

Chu, X. Y., et al. (2019). Improving antitumor outcomes for palliative intratumoral injection therapy through lecithin- chitosan nanoparticles loading paclitaxel- cholesterol complex. *International Journal of Nanomedicine, 14*, 689–705. https://doi.org/10.2147/IJN.S188667

Davis, M. E. (2009). The first targeted delivery of siRNA in humans *via* a self-assembling, cyclodextrin polymer-based nanoparticle: From concept to clinic. *Molecular Pharmaceutics, 6*(3), 659–668. https://doi.org/10.1021/mp900015y

Devalapally, H., Chakilam, A., & Amiji, M. M. (2007). Role of nanotechnology in pharmaceutical product development. *Journal of Pharmaceutical Sciences, 96*(10), 2547–2565. https://doi.org/10.1002/jps.20875

Dilnawaz, F., Acharya, S., & Sahoo, S. K. (2018). Recent trends of nanomedicinal approaches in clinics. *International Journal of Pharmaceutics, 538*(1), 263–278. https://doi.org/10.1016/j.ijpharm.2018.01.016

Dobry, D. E., et al. (2009). A model-based methodology for spray-drying process development. *Journal of Pharmaceutical Innovation, 4*(3), 133–142. https://doi.org/10.1007/s12247-009-9064-4

Duncan, R., & Izzo, L. (2005). Dendrimer biocompatibility and toxicity. *Advanced Drug Delivery Reviews, 57*(15), 2215–2237. https://doi.org/10.1016/j.addr.2005.09.019

Duong, V. A., et al. (2019). Preparation of ondansetron hydrochloride-loaded nanostructured lipid carriers using solvent injection method for enhancement of pharmacokinetic properties. *Pharmaceutical Research, 36*(10). https://doi.org/10.1007/S11095-019-2672-X

Etheridge, M. L., et al. (2013). 'The big picture on nanomedicine: The state of investigational and approved nanomedicine products. *Nanomedicine: Nanotechnology, Biology and Medicine, 9*(1), 1–14. https://doi.org/10.1016/j.nano.2012.05.013

Fang, C.-L., Al-Suwayeh, S. A., & Fang, J.-Y. (2013). Nanostructured lipid carriers (NLCs) for drug delivery and targeting. *Recent Patents on Nanotechnology, 7*(1), 41–55. https://doi.org/10.34172/apb.2020.021

Fischer, H. C., & Chan, W. C. W. (2007). Nanotoxicity: The growing need for in vivo study. *Current Opinion in Biotechnology, 18*(6), 565–571. https://doi.org/10.1016/j.copbio.2007.11.008

Ganesan, P., & Narayanasamy, D. (2017). Lipid nanoparticles: Different preparation techniques, characterization, hurdles, and strategies for the production of solid lipid nanoparticles and nanostructured lipid carriers for oral drug delivery. *Sustainable Chemistry and Pharmacy, 6*, 37–56. https://doi.org/10.1016/j.scp.2017.07.002

Gao, L., et al. (2013). Application of drug nanocrystal technologies on oral drug delivery of poorly soluble drugs. *Pharmaceutical Research, 30*(2), 307–324. https://doi.org/10.1007/s11095-012-0889-z

Guimarães, D., et al. (2020). Increased encapsulation efficiency of methotrexate in liposomes for rheumatoid arthritis therapy. *Biomedicine, 8*(12), 630. https://doi.org/10.3390/biomedicines8120630

Guinebretière, S., et al. (2002). Nanocapsules of biodegradable polymers: Preparation and characterization by direct high resolution electron microscopy. *Materials Science and Engineering: C, 21*(1), 137–142. https://doi.org/10.1016/S0928-4931(02)00073-5

Haider, T., et al. (2022). Nisin and nisin-loaded nanoparticles: A cytotoxicity investigation. *Drug Development and Industrial Pharmacy, 48*(7), 310–321. https://doi.org/10.1080/03639045.2022.2111438

Hammad, M. A., & Müller, B. W. (1998). Solubility and stability of tetrazepam in mixed micelles. *European Journal of Pharmaceutical Sciences, 7*(1), 49–55. https://doi.org/10.1016/S0928-0987(98)00006-2

Hanafy, A., et al. (2007). Pharmacokinetic evaluation of oral fenofibrate nanosuspensions and SLN in comparison to conventional suspensions of micronized drug. *Advanced Drug Delivery Reviews, 59*(6), 419–426. https://doi.org/10.1016/j.addr.2007.04.005

Harsha, S. N., et al. (2015). Nanoparticle formulation by Büchi B-90 nano spray dryer for oral mucoadhesion. *Drug Design, Development and Therapy, 9*, 273–282. https://doi.org/10.2147/DDDT.S66654

Hrkach, J., et al. (2012). Preclinical development and clinical translation of a PSMA-targeted docetaxel nanoparticle with a differentiated pharmacological profile. *Science Translational Medicine, 4*(128), 128ra39. https://doi.org/10.1126/scitranslmed.3003651

Jaiswal, P., Gidwani, B., & Vyas, A. (2016). Nanostructured lipid carriers and their current application in targeted drug delivery. *Artificial Cells, Nanomedicine, and Biotechnology, 44*(1), 27–40. https://doi.org/10.3109/21691401.2014.909822

Jeong, Y.-I., et al. (2008). Ciprofloxacin-encapsulated poly(dl-lactide-co-glycolide) nanoparticles and its antibacterial activity. *International Journal of Pharmaceutics, 352*(1), 317–323. https://doi.org/10.1016/j.ijpharm.2007.11.001

Johnson, K. P., et al. (1998). Extended use of glatiramer acetate (Copaxone) is well tolerated and maintains its clinical effect on multiple sclerosis relapse rate and degree of disability. Copolymer 1 multiple sclerosis study group. *Neurology, 50*(3), 701–708. https://doi.org/10.1212/wnl.50.3.701

Kabanov, A. V., & Alakhov, V. Y. (2002). Pluronic block copolymers in drug delivery: From micellar nanocontainers to biological response modifiers. *Critical Reviews in Therapeutic Drug Carrier Systems, 19*(1), 1–72. https://doi.org/10.1615/critrevtherdrugcarriersyst.v19.i1.10

Khodabandehloo, H., Zahednasab, H., & Ashrafi Hafez, A. (2016). Nanocarriers usage for drug delivery in cancer therapy. *Iranian Journal of Cancer Prevention, 9*(2), e3966. https://doi.org/10.17795/ijcp-3966

Kumar, R., et al. (2022). Lipid based nanocarriers: Production techniques, concepts, and commercialization aspect. *Journal of Drug Delivery Science and Technology, 74*, 103526. https://doi.org/10.1016/j.jddst.2022.103526

Kwon, G. S., & Okano, T. (1996). Polymeric micelles as new drug carriers. *Advanced Drug Delivery Reviews, 21*(2), 107–116. https://doi.org/10.1016/S0169-409X(96)00401-2

Lai, F., et al. (2011). Nanocrystals as tool to improve piroxicam dissolution rate in novel orally disintegrating tablets. *European Journal of Pharmaceutics and Biopharmaceutics: Official Journal of Arbeitsgemeinschaft fur Pharmazeutische Verfahrenstechnik e.V, 79*(3), 552–558. https://doi.org/10.1016/j.ejpb.2011.07.005

Lee, Y. S., et al. (2013). Production of nanoparticles-in-microparticles by a double emulsion method: A comprehensive study. *European Journal of Pharmaceutics and Biopharmaceutics: official journal of Arbeitsgemeinschaft fur Pharmazeutische Verfahrenstechnik e.V, 83*(2), 168–173. https://doi.org/10.1016/J.EJPB.2012.10.016

Leleux, J., & Williams, R. O. (2014). Recent advancements in mechanical reduction methods: Particulate systems. *Drug Development and Industrial Pharmacy, 40*(3), 289–300. https://doi.org/10.3109/03639045.2013.828217

Lertsutthiwong, P., Rojsitthisak, P., & Nimmannit, U. (2009). Preparation of turmeric oil-loaded chitosan-alginate biopolymeric nanocapsules. *Materials Science and Engineering: C, 29*(3), 856–860. https://doi.org/10.1016/j.msec.2008.08.004

Li, M., et al. (2016). Graphene oxide framework: An adsorbent for solid phase extraction of phenylurea herbicides from water and celery samples. *Journal of Chromatography A, 1469*, 17–24. https://doi.org/10.1016/j.chroma.2016.09.056

Li, W., et al. (2017). Preparation of CaP/pDNA nanoparticles by reverse micro-emulsion method: Optimization of formulation variables using experimental design. *Asian Journal of Pharmaceutical Sciences, 12*(2), 179–186. https://doi.org/10.1016/J.AJPS.2016.09.006

Li, X., et al. (2021). Magnetic nanoparticles for cancer theranostics: Advances and prospects. *Journal of Controlled Release: Official Journal of the Controlled Release Society, 335*, 437–448. https://doi.org/10.1016/j.jconrel.2021.05.042

Liu, J., et al. (2010). A modified double-emulsion method for the preparation of daunorubicin-loaded polymeric nanoparticle with enhanced in vitro anti-tumor activity. *Biomedical Materials (Bristol, England), 5*(6). https://doi.org/10.1088/1748-6041/5/6/065002

Liu, C., et al. (2017). Morphology and characteristics of starch nanoparticles self-assembled via a rapid ultrasonication method for peppermint oil encapsulation. *Journal of Agricultural and Food Chemistry, 65*(38), 8363–8373. https://doi.org/10.1021/ACS.JAFC.7B02938

Louey, M. D., Van Oort, M., & Hickey, A. J. (2004). Aerosol dispersion of respirable particles in narrow size distributions produced by jet-milling and spray-drying techniques. *Pharmaceutical Research, 21*, 1200–1206. https://doi.org/10.1023/b:pham.0000033007.27278.60

McKay, P. F., et al. (2020). Self-amplifying RNA SARS-CoV-2 lipid nanoparticle vaccine candidate induces high neutralizing antibody titers in mice. *Nature Communications, 11*(1), 3523. https://doi.org/10.1038/s41467-020-17409-9

Mehnert, W., & Mäder, K. (2001). Solid lipid nanoparticles: Production, characterization and applications. *Advanced Drug Delivery Reviews, 47*(2–3), 165–196. https://doi.org/10.1016/s0169-409x(01)00105-3

Merisko-Liversidge, E., & Liversidge, G. G. (2011). Nanosizing for oral and parenteral drug delivery: A perspective on formulating poorly-water soluble compounds using wet media milling technology. *Advanced Drug Delivery Reviews, 63*(6), 427–440. https://doi.org/10.1016/j.addr.2010.12.007

Mohammady, M., Mohammadi, Y., & Yousefi, G. (2020). Freeze-drying of pharmaceutical and nutraceutical nanoparticles: The effects of formulation and technique parameters on nanoparticles characteristics. *Journal of Pharmaceutical Sciences, 109*(11), 3235–3247. https://doi.org/10.1016/j.xphs.2020.07.015

Möschwitzer, J. P. (2013). Drug nanocrystals in the commercial pharmaceutical development process. *International Journal of Pharmaceutics, 453*(1), 142–156. https://doi.org/10.1016/j.ijpharm.2012.09.034
Möschwitzer, J., & Müller, R. H. (2006). Spray coated pellets as carrier system for mucoadhesive drug nanocrystals. *European Journal of Pharmaceutics and Biopharmaceutics, 62*(3), 282–287. https://doi.org/10.1016/j.ejpb.2005.09.005
Müller, R. H., Jacobs, C., & Kayser, O. (2001). Nanosuspensions as particulate drug formulations in therapy: Rationale for development and what we can expect for the future. *Advanced Drug Delivery Reviews, 47*(1), 3–19. https://doi.org/10.1016/S0169-409X(00)00118-6
Oerlemans, C., et al. (2010). Polymeric micelles in anticancer therapy: Targeting, imaging and triggered release. *Pharmaceutical Research, 27*(12), 2569–2589. https://doi.org/10.1007/s11095-010-0233-4
Onoue, S., et al. (2011). Development of inhalable nanocrystalline solid dispersion of tranilast for airway inflammatory diseases. *Journal of Pharmaceutical Sciences, 100*(2), 622–633. https://doi.org/10.1002/jps.22299
Paliwal, S. R., et al. (2010). A novel cancer targeting approach based on estrone anchored stealth liposome for site-specific breast cancer therapy. *Current Cancer Drug Targets, 10*(3), 343–353. https://doi.org/10.2174/156800910791190210
Pardi, N., et al. (2018). mRNA vaccines – A new era in vaccinology. *Nature Reviews Drug Discovery, 17*(4), 261–279. https://doi.org/10.1038/nrd.2017.243
Patravale, V. B., Date, A. A., & Kulkarni, R. M. (2004). Nanosuspensions: A promising drug delivery strategy. *Journal of Pharmacy and Pharmacology, 56*(7), 827–840. https://doi.org/10.1211/0022357023691
Patton, D. L., et al. (2006). Preclinical safety and efficacy assessments of dendrimer-based (SPL7013) microbicide gel formulations in a nonhuman primate model. *Antimicrobial Agents and Chemotherapy, 50*(5), 1696–1700. https://doi.org/10.1128/AAC.50.5.1696-1700.2006
Pedersen, N., et al. (2006). Solid lipid nanoparticles can effectively bind DNA, streptavidin and biotinylated ligands. *European Journal of Pharmaceutics and Biopharmaceutics : Official Journal of Arbeitsgemeinschaft fur Pharmazeutische Verfahrenstechnik e.V, 62*(2), 155–162. https://doi.org/10.1016/j.ejpb.2005.09.003
Rabinow, B. E. (2004). Nanosuspensions in drug delivery. *Nature Reviews Drug Discovery, 3*(9), 785–796. https://doi.org/10.1038/nrd1494
Rasenack, N., & Müller, B. W. (2004). Micron-size drug particles: Common and novel micronization techniques. *Pharmaceutical Development and Technology, 9*(1), 1–13. https://doi.org/10.1081/pdt-120027417
Rautio, J., et al. (2008). Prodrugs: Design and clinical applications. *Nature Reviews Drug Discovery, 7*(3), 255–270. https://doi.org/10.1038/nrd2468
Rijcken, C. J. F., et al. (2005). Novel fast degradable thermosensitive polymeric micelles based on PEG-block-poly(N-(2-hydroxyethyl)methacrylamide-oligolactates). *Biomacromolecules, 6*(4), 2343–2351. https://doi.org/10.1021/bm0502720
Salazar, J., Müller, R. H., & Möschwitzer, J. P. (2014). Combinative particle size reduction technologies for the production of drug nanocrystals. *Journal of Pharmaceutics, 2014*(1–14). https://doi.org/10.1155/2014/265754
Schmid, K. E., et al. (2006). Update on ocular complications of systemic cancer chemotherapy. *Survey of Ophthalmology, 51*(1), 19–40. https://doi.org/10.1016/j.survophthal.2005.11.001
Schubert, M. A., & Müller-Goymann, C. C. (2003). Solvent injection as a new approach for manufacturing lipid nanoparticles–evaluation of the method and process parameters. *European Journal of Pharmaceutics and Biopharmaceutics, 55*(1), 125–131. https://doi.org/10.1016/s0939-6411(02)00130-3
Shah, R., et al. (2015). *Lipid nanoparticles: Production, characterization and stability*. Springer. https://doi.org/10.1007/978-3-319-10711-0
Siepmann, J., & Siepmann, F. (2013). Mathematical modeling of drug dissolution. *International Journal of Pharmaceutics, 453*(1), 12–24. https://doi.org/10.1016/j.ijpharm.2013.04.044

Singh, M., et al. (2021). Streptomycin sulphate loaded solid lipid nanoparticles show enhanced uptake in macrophage, lower MIC in mycobacterium and improved oral bioavailability. *European Journal of Pharmaceutics and Biopharmaceutics: Official Journal of Arbeitsgemeinschaft fur Pharmazeutische Verfahrenstechnik e.V, 160*, 100–124. https://doi.org/10.1016/J.EJPB.2021.01.009

Sinha, B., Müller, R. H., & Möschwitzer, J. P. (2013). Bottom-up approaches for preparing drug nanocrystals: Formulations and factors affecting particle size. *International Journal of Pharmaceutics, 453*(1), 126–141. https://doi.org/10.1016/j.ijpharm.2013.01.019

Soto-Castro, D., et al. (2012). Solubilization and anticancer-activity enhancement of methotrexate by novel dendrimeric nanodevices synthesized in one-step reaction. *Bioorganic Chemistry, 41–42*, 13–21. https://doi.org/10.1016/j.bioorg.2012.01.002

Svilenov, H., & Tzachev, C. (2014). Solid lipid nanoparticles—A promising drug delivery system. *Nanomedicine, 2*, 187–237. https://doi.org/10.2174/138161282566619090315532l

Swanson, D. R., et al. (2007). Unique steric and geometry induced stoichiometries observed in the divergent synthesis of poly(ester-acrylate/amine) (PEA) dendrimers. *New Journal of Chemistry, 31*(7), 1368–1378. https://doi.org/10.1039/B700193M

Talluri, S. V., et al. (2016). Lipid-based nanocarriers for breast cancer treatment – Comprehensive review. *Drug Delivery, 23*(4), 1291–1305. https://doi.org/10.3109/10717544.2015.1092183

Tan, E. H., et al. (2017). Downstream drug product processing of itraconazole nanosuspension: Factors influencing tablet material properties and dissolution of compacted nanosuspension-layered sugar beads. *International Journal of Pharmaceutics, 532*(1), 131–138. https://doi.org/10.1016/j.ijpharm.2017.08.107

Tomalia, D. A., Naylor, A. M., & Goddard, W. A., III. (1990). Starburst dendrimers: Molecular-level control of size, shape, surface chemistry, topology, and flexibility from atoms to macroscopic matter. *Angewandte Chemie International Edition in English, 29*(2), 138–175. https://doi.org/10.1002/anie.199001381

Torchilin, V. P. (2005). Recent advances with liposomes as pharmaceutical carriers. *Nature Reviews Drug Discovery, 4*(2), 145–160. https://doi.org/10.1038/nrd1632

Vaage, J., et al. (1992). Therapy of primary and metastatic mouse mammary carcinomas with doxorubicin encapsulated in long circulating liposomes. *International Journal of Cancer, 51*(6), 942–948. https://doi.org/10.1002/ijc.2910510618

Van Eerdenbrugh, B., et al. (2007). Characterization of physico-chemical properties and pharmaceutical performance of sucrose co-freeze-dried solid nanoparticulate powders of the anti-HIV agent loviride prepared by media milling. *International Journal of Pharmaceutics, 338*(1–2), 198–206. https://doi.org/10.1016/j.ijpharm.2007.02.005

Veiseh, O., Gunn, J. W., & Zhang, M. (2010). Design and fabrication of magnetic nanoparticles for targeted drug delivery and imaging. *Advanced Drug Delivery Reviews, 62*(3), 284–304. https://doi.org/10.1016/j.addr.2009.11.002

Verma, V., Ryan, K. M., & Padrela, L. (2021). Production and isolation of pharmaceutical drug nanoparticles. *International Journal of Pharmaceutics, 603*, 120708. https://doi.org/10.1016/j.ijpharm.2021.120708

Wang, C., et al. (2022). Construction and characterization of novel hydrophilic nanospheres loaded with lambda-Cyhalothrin via ultrasonic emulsification-solvent evaporation. *International Journal of Molecular Sciences, 23*(22). https://doi.org/10.3390/IJMS232214063

Wu, X., et al. (2009). Optimisation of Dex-GMA nanoparticles prepared in modified micro-emulsion system: Physical and biologic characterization. *Journal of Biotechnology, 143*(4), 268–273. https://doi.org/10.1016/J.JBIOTEC.2009.07.015

Xie, M., Chen, Y., & Wu, L. (2013). Preparation of doxorubicin-hydrochloride nanoliposomes by ethanol injection-pH gradient method and their safety evaluation. *Journal of Nanoscience and Nanotechnology, 13*(1), 216–221. https://doi.org/10.1166/JNN.2013.6798

Yallapu, M. M., et al. (2010). PEG-functionalized magnetic nanoparticles for drug delivery and magnetic resonance imaging applications. *Pharmaceutical Research, 27*(11), 2283–2295. https://doi.org/10.1007/s11095-010-0260-1

You, J., et al. (2007). Preparation and characteristic of vinorelbine bitartrate-loaded solid lipid nanoparticles. *International Journal of Pharmaceutics, 343*(1–2), 270–276. https://doi.org/10.1016/J.IJPHARM.2007.07.003

Zhang, H., et al. (2013). Novel formation of ag/au bimetallic nanoparticles by physical mixture of monometallic nanoparticles in dispersions and their application to catalysts for aerobic glucose oxidation. *Langmuir, 29*(33), 10330–10339. https://doi.org/10.1021/la401878g

Zhang, X., et al. (2021). Rice peptide nanoparticle as a bifunctional food-grade Pickering stabilizer prepared by ultrasonication: Structural characteristics, antioxidant activity, and emulsifying properties. *Food Chemistry, 343*. https://doi.org/10.1016/J.FOODCHEM.2020.128545

Zhu, Y., et al. (2005). Influence of surfactants on the parameters of polylactide nanocapsules containing insulin. *Journal of Surfactants and Detergents, 8*(4), 353–358. https://doi.org/10.1007/s11743-005-0368-9

Zili, Z., Sfar, S., & Fessi, H. (2005). Preparation and characterization of poly-ε-caprolactone nanoparticles containing griseofulvin. *International Journal of Pharmaceutics, 294*(1), 261–267. https://doi.org/10.1016/j.ijpharm.2005.01.020

Zimmermann, C. M., et al. (2022). Spray drying siRNA-lipid nanoparticles for dry powder pulmonary delivery. *Journal of Controlled Release: Official Journal of the Controlled Release Society, 351*, 137–150. https://doi.org/10.1016/J.JCONREL.2022.09.021

Chapter 12
Scale-Up of Nanoparticle Manufacturing Process

Clara Fernandes, Manasi Jathar, Bhakti Kubal Shweta Sawant, and Tanvi Warde

Contents

Abstract Nanoparticles owning a particle size of less than 100 nm have versatile use in the pharmaceutical industry. Different nanoparticles have been produced depending upon the material, viz. gold, silver, carbon, polymer, etc. The role of nanoparticles includes drug carriers for both hydrophilic and lipophilic drugs, targeting, enhancing bioavailability with reduced toxicity, imaging, biosensing, sustained/controlled release, etc. Modification of nanoparticles with capping agents provides an add-on benefit in targeting and preventing aggregation. Nevertheless, the scalability of nanoparticles with uniform size, distribution, and shape is a major challenge. Broadly, the production of nanoparticles can be classified as top-down and bottom-up methods. Top-down approaches consist of energy-intensive methods to reduce the bulk into nanosized particles. Whereas bottom-up approaches aim to collate atoms to generate nanosized particles. Although nanoparticles provide various advantages, industrial production is difficult due to not well-defined optimization parameters to provide reproducible nanoparticles. The problems faced during industrial production include aggregation, contamination, degradation of nanoparticles along with low yield questioning the economic aspect of scalability. These lead to hindrances in the industry acceptability of nanoparticle production. The challenges in scale-up production and the developed methods to tackle these problems are reviewed in the book chapter.

Keywords Scaleup · Nanoparticle · Top-down · Bottom-up · Process parameter

C. Fernandes (✉) · M. Jathar · B. K. S. Sawant · T. Warde
IPA-MSB's Bombay College of Pharmacy, Mumbai, Maharashtra, India
e-mail: clara.fernandes@bcp.edu.in

A. B. Jindal (ed.), *Pharmaceutical Process Engineering and Scale-up Principles*, AAPS Introductions in the Pharmaceutical Sciences 13,
https://doi.org/10.1007/978-3-031-31380-6_12

12.1 Introduction

The pharmaceutical industry is witnessing an impetus in the market- and technology-driven delivery options with the sole intent to improvize the therapeutic efficacy of drugs as well as expand the market share of the existing product through brand differentiation (Fig. 12.1). The emerging trends reflect the potential of innovative nano-delivery options to achieve targeted delivery with reduced side effects at reasonable therapy costs. However, scalability, an essential pre-requisite for product commercialization, remains an unresolved challenge for nanotechnology in the pharmaceutical arena (*Process Development & Scale-up*). Several factors are known to contribute to the issue mentioned above, the most important being a selection of suitable production technology for nanoparticles. In literature, there are a plethora of technologies cited for lab-scale production of nanoparticles irrespective of the drug or excipient. Unlike the macro dosage form, there is a lack of clarity on the choice of technology that can be adopted. For instance, a drug such as cyclosporine exists in various types of formulations such as oral solutions (Panimun Bioral®) capsules (Psorid 100®), intravenous injection (Cyclosporine injection USP), eye drops (Hydroeyes®), hence creating hurdles in deciding for choosing the technology for nanopreparations.

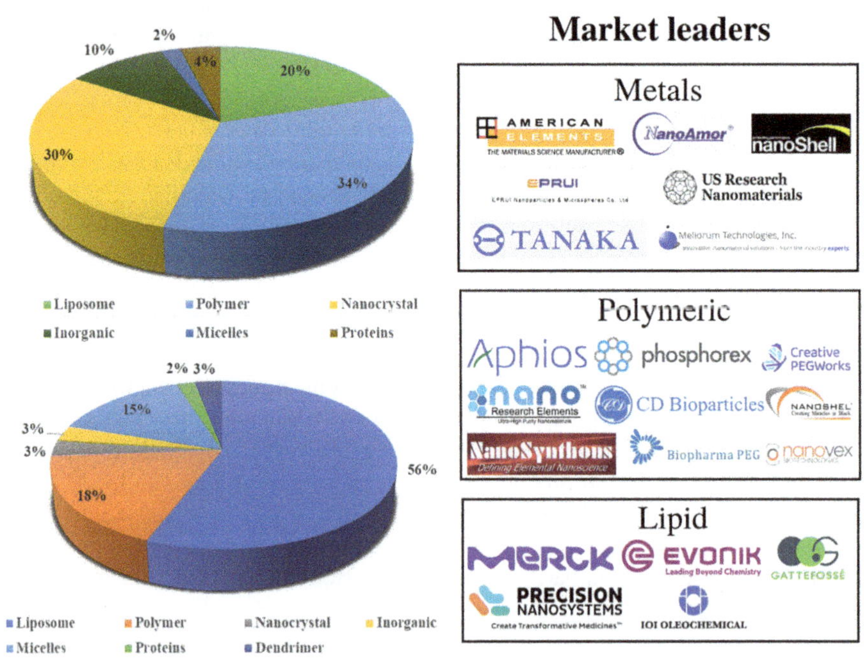

Fig. 12.1 Overview of nanoparticles present in research and clinical use and the prominent market leaders. (Adapted from 'Large scale manufacturing of nanoparticles-An industrial outlook'). (**a**) Nanoparticles approved by regulatory authorities for clinical use (**b**) Nanoparticles approved by regulatory authorities for investigational drugs

Moreover, from a regulatory perspective, there is a dearth of lab technology which can conform to regulatory guidelines to be GMP certified (ICH guidelines Q9), qualified and validated for obtaining approval from regulatory agencies such as the FDA (Shegokar & Nakach, 2020). Industry acceptance of nanoparticles also depends upon the cost of the finished product. Thus, cost needs to be considered as one of the essential factors. If the equipment selected is not appropriate, the issues that are faced during production will lead to an increased workload on the equipment leading to damage to the equipment. This will add to the maintenance cost other than the existing installation cost. This will further add to the production cost and thus increase the selling cost of the finished product. There is a pricing cap below which the finished products will be sold. Hence, the major bottleneck faced by nanotechnology in its translation from lab to commercial scale is the lack of adaptability of the standard technologies as it demands the use of specialized and costly equipment (such as high-pressure homogenizer, planetary ball mill, etc.), incurring high capital investment. Furthermore, variability is often encountered in batch-to-batch production of nanoparticles due to the inherent stability formulation issues (agglomeration, aggregation), and undefined characterization tools. The propensity of the altered toxicity profile of the nanoformulation distinct from excipients/ drugs renders the risk assessment during manufacturing a daunting task (*Process Development & Scale-up – www.vbpgroup-ict.in*, n.d.; Soares et al., 2018; Liu & Meng, 2021).

Nonetheless, it is forecasted that the nanomedicine market will be USD 964.15 Billion by the year 2030 indicating the potential of nanotechnology in changing the healthcare sector. Some of the existing marketed nanoparticle-containing formulation is as mentioned in Table 12.1.

It is evident from the table, although the market share is low, there is a considerable presence of nanoformulation. Considering the advantages of nanoformulations, this book chapter will provide an overview of the parameters which need to be considered for the scale-up of industry-feasible technologies.

12.2 Classification of Techniques for Preparation of Nanoparticles

Nanoparticle manufacturing can be broadly classified into two approaches: top-down and bottom-up (Fig. 12.2). The top-down approach is achieved using different mechanical (milling), cavitation techniques (sonication, High-Pressure Homogenization) and thermal techniques (thermal decomposition methods). Similarly, the bottom-up approach is usually accomplished by processes such as thermal techniques (chemical vapor deposition (CVD), laser pyrolysis), precipitation, and sol-gel processing (Shegokar & Nakach, 2020; Subhan et al., 2021). These approaches can further be classified into physical, chemical, and biological methods based upon means of process accompanied by the synthesis of nanoparticles.

Table 12.1 Marketed preparations of nanoparticle-containing formulations (Soares et al., 2018)

Sr. no.	Drug	Therapeutic use	Nanomedicine class	Brand name	Manufacturing company	References
1.	Nab-paclitaxel	Breast neoplasms	Nanoparticles	Abraxane®	Bristol Myers Squibb	*ABRAXANE® – Official Patient & Caregiver Website* (n.d.)
2.	Glatiramer acetate	Multiple sclerosis	Nanoparticles	Copaxone®	Teva	*COPAXONE® ǀ Medication to Treat Relapsing Multiple Sclerosis (RMS)* (n.d.)
3.	Daunorubicin	Advanced HIV-related Kaposi's Sarcoma	Liposomes	Daunoxome®	Gilead Sciences, Inc. (now Galen)	*Galen acquires DaunoXome® (daunorubicin citrate liposome injection) for advanced HIV-associated Kaposi's sarcoma ǀ Fierce Biotech* (n.d.)
4.	Amphotericin B	Fungal infection	Liposomes	Ambisome®	Gilead Sciences, Inc.	*AmBisome® ǀ AmBisome (amphotericin B) liposome for injection* (n.d.)
5.	Ferric carboxymaltose	Iron deficiency	Nanocomplex	Ferinject®	Vifor Pharma	*Ferinject (ferric carboxymaltose) – Summary of Product Characteristics (SmPC) – (emc)* (n.d.)
6.	Verteporfin	Age-related macular degeneration	Nanocomplex	Visudyne®	Bausch + Lomb	*XIPERE®, VISUDYNE® and RETISERT® for Retinal Condition Management: Bausch + Lomb* (n.d.)
7.	Cyclosporine	Rheumatoid arthritis, atopic dermatitis, bone marrow transplantation	Nanoemulsion	Sandimmun Neoral®	Novartis	*Sandimmune (Cyclosporine): Uses, Dosage, Side Effects, Interactions, Warning* (n.d.)
8.	Pegaspargase	Acute lymphoblastic leukaemia	Nanoemulsion	Oncaspar®	Servier Pharmaceuticals LLC	*ONCASPAR – Use in Treating Acute Lymphoblastic Leukemia (ALL)* (n.d.)
9.	Fenofibrate	Hyperlipidaemia	Nanocrystal	Tricor®	AbbVie Inc.	*Tricor for Cholesterol: Uses, Dosage, Side Effects, Interactions, Warnings* (n.d.)
10.	Sirolimus	Graft rejection	Nanocrystal	Rapamune®	Pfizer	*RAPAMUNE® ǀ Pfizer* (n.d.)

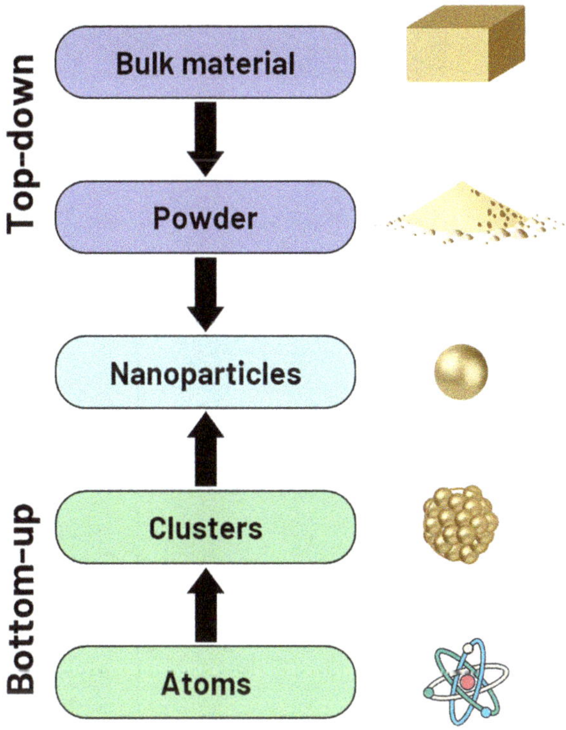

Fig. 12.2 Brief overview of the top-down and bottom-up approach

The physical method includes ball milling, electric arc discharge method, laser ablation method, and physical vapour deposition method. The advantage of synthesizing nanoparticles by the physical process is that it provides uniform particle size distribution and high purity. However, it has limitations of being complex equipment to use and needs external energy. The physical method can produce nanoparticles even on a large scale. However, there is a possibility of the formation of aggregates leading to the formation of large-sized particles.

Chemical methods, including co-precipitation, sonochemical method (ultrasonication), gas condensation method, and microwave synthesis, are generally employed to process the liquid and gas phase production of nanoparticles. Biological methods include the preparation of nanoparticles either by extraction from plant origin such as orange peel or with the assistance of microorganisms such as bacteria, fungi, and algae (Subhan et al., 2021). A more comprehensive classification of techniques employed for the preparation of nanoparticles can be depicted as given in the figure below:

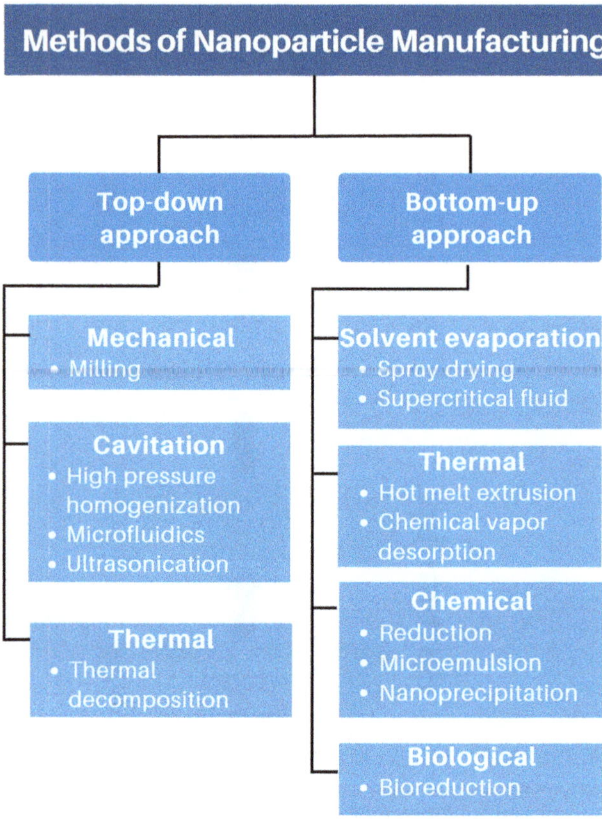

Fig. 12.3 Classification of techniques for manufacturing nanoparticle

12.2.1 Top-Down Approach

The top-down approach is where macro-sized particles are broken down into nanosized particles. Based upon the mechanism of nanosizing (Fig. 12.3), it can be further classified into three categories:

1. Mechanical techniques (milling)
2. Cavitation techniques
   (a) Ultrasonication
   (b) High-pressure homogenization
   (c) Microfluidics technique
3. Thermal Techniques
   (a) Chemical vapour deposition
   (b) Microwave-assisted thermal decomposition
   (c) Laser ablation technique

12.2.1.1 Mechanical Techniques

Particles, when subjected to mechanical energy, are physically crushed to sizes smaller than the original. Techniques such as mechanical grinding, milling, and alloying follow the above principle and are catalogued under mechanical techniques. Milling, amongst the techniques mentioned above, is widely used. Ball mill, colloid mill, hammer mill, fluidized energy mill, roller mill, cone mill, etc. are the various types of mills that are utilized for particle size reduction in the pharmaceutical industry. However, only the ball mill has been indicated to be of assistance in the preparation of nanoparticles. Hence, ball milling is discussed in brief in the following chapter (Loh et al., 2014).

1. *Milling*

The external and internal morphology of nanoparticles contributes towards the numerous application that can be achieved by altering the mechanical and physical characteristics of the material. Milling leads to the amorphization of crystalline structures that can be targeted to enhance the dissolution rate and solubility of the drug. Kobayashi et al. studied the enhanced dissolution rate and solubility effect of atorvastatin calcium sesquihydrate due to the amorphization by ball milling(Kobayashi et al., 2017). Milling is the solid-state process of reducing the size of the particles from coarse particle size to defined particle size. Milling can also be used for synthesis of nanoparticles as first developed by John Benjamin (1970) utilizing high energy ball mill. Mechanical milling is the process of ball-powder-ball collision. which leads to formation of fractures in the feed material due to conversion of stored energy (potential energy) into free surface energy (*Nanoparticles from Mechanical Attrition*, n.d.). The figure (Fig. 12.4) represents the steps included in the mechanical compaction as a part of milling process. It starts with minor elastic and plastic deformation of the particles which is followed by rearrangement of the particles in the created spaces. The particles develop fracture in the final stage i.e., fragmentation (*Nanoparticles from Mechanical Attrition*).

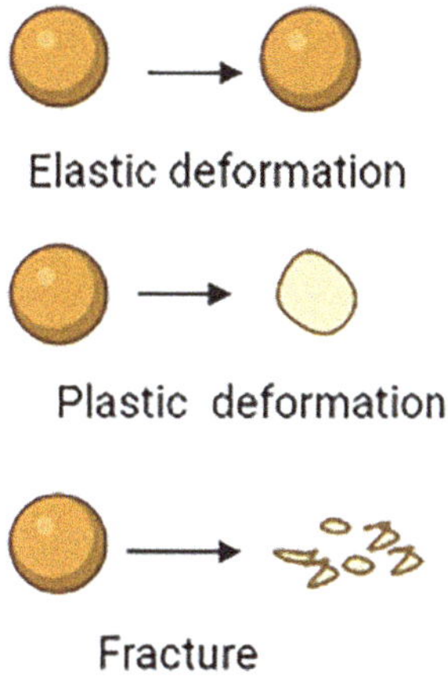

Fig. 12.4 Deformation during milling (**a**) Elastic deformation (**b**) Plastic deformation (**c**) Fracture

This results in break down of chemical bonds present in the material causing generation of enormous kinetic energy leading to size reduction (Kumar et al., 2020). They may follow crushing, impact, and attrition forces to dissipate this energy. The mechanism of milling can be explained by Rittinger's, Kick's, and Bond's theory for the energy requirement for size reduction (Wills & Finch, 2016).

Rittinger's Theory (P. R. Rittinger, "Lehrbuch der Aufbereitungskunde," Berlin, 1867. – References – Scientific Research Publishing, n.d.)

It represents the energy consumed in size reduction is proportional to the area of new surface produced. The surface area of a known weight of particles of uniform diameter is inversely proportional to the diameter, and it is expressed as follows:

$$E_{VR} = K\left(1/_P - 1/_F\right)$$

Here, E_{VR} = energy, K = constant, P = Initial surface area, F = New surface area.

Kick's Theory

It represents the energy required is proportional to the reduction in the volume of the particles, and it is expressed as follows:

$$E_K = K\left(\mathrm{In}^F/_P\right)$$

Here, K = Constant, F/P = reduction ratio.

Bond's Theory

It represents the energy input is proportional to the new crack tip length produced in particle breakage, and it is expressed as follows:

$$W = 10 \times W_i\left(1/\sqrt{P_{80}} - 1/\sqrt{F_{80}}\right)$$

Here, W = energy input (work) in kilowatt hours per metric ton, W_i = work index P_{80} and F_{80} are the 80% product and feed passing sizes, in micrometres.

Based on the nature of the material, different mechanical behaviour of the produced particles will be established (i.e., ductile-ductile, ductile-brittle, and brittle-brittle). It is initiated with the process of micro-forging, i.e., flattening of ductile powders into plates and fragments and further accompanied by cold welding leading to finer particles formation. In case of ductile-brittle material they are entrapped among the free spaces between the particles to form a powder blend. Whereas, in case of brittle-brittle material, it undergoes simple size reduction, also called the limit of combination. This means that the particle reduction takes place plastically, not by deformation or fracturing the particles (Prasad Yadav et al., 2012) (i.e., explained by Griffith's theory that represent the stress at which crack initiation happens). The macro-sized particle formation can be done using different mills such as ball mills, hammer mills, colloid mills, fluid energy mills, and roller compression mills. But, for nanoparticles, it is best suited using ball mills based on energy requirements which are further classified as mentioned in Table 12.2 (Sherif El-Eskandarany et al., 2021).

Process parameters must be considered while using the ball mill to prepare these nanoparticles (Fig. 12.5). They also need to be considered when the process is scaled up. These are as follows:

A. *Design Characteristics* (Wills & Finch, 2016) Type of mill to be used (which is decided based on the milling goals and powder type) as the balls, raw materials and reagents are mixed while rotating the vessel in a specified direction, causing impact or pulling forces to the material (Howard et al., 2018).

The material of the milling media influences the size of particles. The hardness and large surface area are the key points to be taken into consideration. There are several materials of choice for the milling media, like hardened steel, tungsten carbide, agate, and zirconia. These are essential to maintain the milling efficiency and decreases the milling time for obtaining homogenous powder particles. An important characteristics that needs to be considered while choosing the milling media is the hardness of the powder material. The hardness of powder material is considered to be the most important characteristic while deciding on what type of milling media to choose. The harder the milling media, the better the milling efficiency. Calderón Bedoya et al. conducted a study on the influence of vial material (stainless steel and zirconia) in the generation of iron oxide nanoparticles using a high-energy planetary ball mill. It was reported that the synthesis of Iron nanoparticles were strongly influenced by the presence of the starting material and the nature of the container used during milling process. Another study by Bitterlich et al. (2014) reported the comparison of the influence of zirconia and alumina beads of similar size on the milling time. It was observed that the use of zirconia beads resulted in better milling efficiency as an effect of higher energy generation per collision(due to higher density of zirconia beads).

On the contrary, hard milling media also can contaminate the milled powders, hence precautions need to be taken during the processing. Balls will offer a larger surface area than rods, but rods have a greater contact area with no chance of contamination. Thus, according to the end product, selection of the shape of milling media is done.

Ball to powder ratio Increasing the Wb:Wp ratio (Ball – to powder ratio (BPR)) increases the kinetic energy with an enhanced effect on the rate of amorphization. Waje et al. studied the effect of (BPR) on the nanosized cobalt ferrite spinel particles formation by varying the BPR as 8:1, 10:1, 15:1, 20:1, 30:1. The result indicated that the particle size decreases linearly from 15.3 to 11.4 nm when BPR is 8:1 and 30:1, respectively (Waje et al., 2010).

The volume of filling It modulates the trajectory of balls that had been investigated for the Knoevenagel condensation of vanillin with barbituric acid in a planetary mill by Stolle et al. (Howard et al., 2018).

Table 12.2 Overview of different ball mills

Sr. no.	Type of mill	Particle size	Mechanism	Starting material	Process parameters	Grinding media	Advantages	Disadvantages
1.	High energy mill							
	(a) Attritor ball mill	>2 nm	Impact and shear	Fibrous and polymer	Ball velocity and rpm	Aluminium oxide, steel, tungsten	Low power consumption, compact design, small plant area	Limited size (<13 mm)
	(b) Shaker mill (SPEX CertPrep, Inc., Spex8000M and Spex8000D)	>105 μm	Vibration	Brittle	Ball velocity, rpm	Steel, tungsten, zirconia	–	–
	(c) Planetary ball mill (Pulverisette P5)	<1 μm	Impact forces	Hard, soft, brittle, moist	Ball velocity, rpm	Zirconium oxide, stainless steel, tungsten	All types of materials can be used	Wear of container and balls, contamination of the product
2.	Low energy mill							
	(a) Tumbler ball mill	–	Impact, shear, and aberration	Hard material	Ball velocity	Carbide, zirconia, silicon nitride	Uniform powder, high efficiency for dry and wet type	Longer milling time

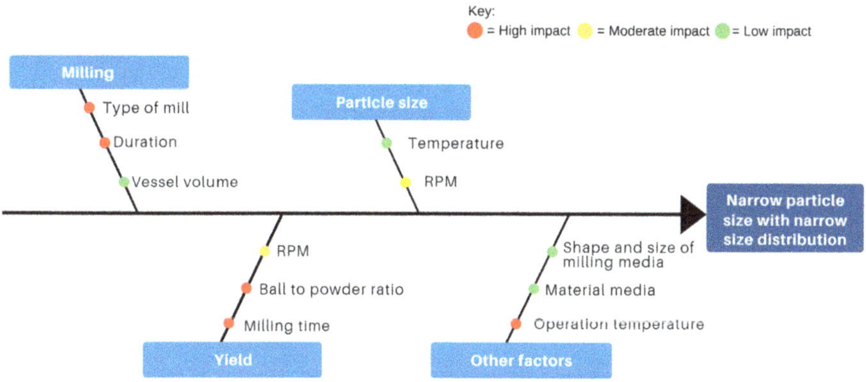

Fig. 12.5 Fishbone diagram summarizing the process parameters affecting the quality attributes milling process. Colour dot indicates the level of risk assessment

B. *Operational characteristics* (Wills & Finch, 2016).

Milling time A longer milling time is associated with more bead–bead collisions and drug particle compressions, which increases the extent of particle breakage and reduces the fraction of un-milled particles. Aldo E. Mariño-Gámez et Al. studied the influence of high-energy ball milling on the structural properties of TiO_2 by modifying the nanoparticle. The samples were extracted at different milling times. The SEM results indicated that, as the milling time increased, the size of nanoparticles became more homogenous and narrower (Mariño-Gámez et al., 2022).

Milling speed, or the rate of circulation, is defined as the number of circulations of the mill a particle makes per mill revolution. Higher milling speed was observed in case of SPEX milling as compared to planetary mill as studied by Zolriasatein et al,. They reported increase in milling energy and smaller particle size was obtained. On comparison between SPEX ball mill and planetary ball mill, for the period of 2 hours, it was observed that the particle size was reduced down to nano size by SPEX and to micro size for Planetary ball mill (Zolriasatein et al., 2018). Milling the powders at high speeds increases the contamination by the materials of the milling tools due to the high rate of wear.

The operating temperature is the most important parameter that needs to be controlled. This becomes the drawback of ball milling as it restricts the type of powder used. Thermolabile powders or extremely soft materials such as polymers or certain materials cannot be used for the preparation of nanoparticles by the ball milling technique. This drawback can be overcome by the recent advances in the ball milling technique, which includes production using cryo-ball mills. These grind the powder to a finer size under the flow of liquid nitrogen. While using this technique, the temperature goes down to −196 °C (Sherif El-Eskandarany et al., 2021).

The milling atmosphere refers to the gas atmosphere inside the milling vial. The process can be run in conditions where in the atmosphere is purged using inert gas (argon, helium) or under normal air conditions. It is essential to consider milling atmosphere as an important factor as presence of oxygen can lead to serious problems in milling process.

PCA (process control agent) is the organic compound used in the milling, acting as the liquid lubricant. Their role during the milling process is to prevent the agglomeration of powders. They are also referred to as particle agglomeration-inhibitor or negative milling. Some of the examples of such PCA are ethyl alcohol, ethanol, etc., but they can also be in solid forms, such as oxalic acid, boric acid, and alumina. They should be subject to drying after the milling process. But, in some cases, high temperature during the processing causes degradation. Thus, the high moisture-containing milled powders should be stored in vacuum-silica gel dissectors to prevent the moisture absorption.

C. *Environmental factors* (Wills & Finch, 2016): Gas contamination due to presence of gases such as oxygen, nitrogen is very common type of environmental parameter that generally affect the unsealed milling vial. This may cause the content to undergo oxidation to form a metal oxide readily, and this reaction is irreversible that cannot be changed by simple drying. Hence, to minimize such contamination, the material handled (raw and milled), mixed and sealed into the vial are surrounded with inert gas environment of argon or helium. Special care has to be taken for the material that are sensitive to Nitrogen like Ti, Zr, and Hf that further, react to give respective nitride in the end product.

12.2.1.2 Cavitation-Based Techniques

Cavitation techniques make use of energy which is generally in the form of sound waves. These waves, when generated at a high amplitude and specific frequency, yield cavitation bubbles which enclose forces strong enough to deform and destroy the existing droplets present. This further lead to small droplet size, hence forming nanosized preparations. Since this technique uses energy as the medium to prepare nanoparticles, cavitation-based techniques are termed green technologies. This technique is generally employed for the preparation of nanoemulsions, a promising technique in the production of uniform nanoemulsions(Sivakumar et al., 2014).

1. *High-Pressure Homogenization*

High-pressure homogenizer is commonly used to prepare nanosuspensions wherein the suspension is driven at a higher flow rate and a pressure above 20,000 psi. A high-pressure homogenizer construction includes an interaction chamber and a high-pressure pump. The pump drives the suspension through the interaction chamber at supersonic speed through a small orifice, where the suspension is subjected to mechanical forces such as high shear (caused due to presence of a gap between the fluid and the valve leading to interruption of the movement of the fluid) (Martínez-Monteagudo et al., 2016) and turbulence. Due to the high flow rate and pressure, there is a change in temperature and vapour pressure that is observed in the process. These changes in the process lead to the formation and collapse of cavitation bubbles, subjecting to increased force on the droplets, leading to the formation of nanosized droplets. However, for achieving uniform and fine droplet-sized suspension/ emulsion, pre-homogenization is essential (Loh et al., 2014). High-pressure homogenization usually achieves narrow particle size distribution and change of properties of the drug such as alteration in shape, size as well as chemical properties (amorphization).

Process parameters that need to be considered while using high-pressure homogenizer are as follows (Fig. 12.6):

A. *Geometry of the valve* Valve is intended to interrupt the flow of the fluids and cause turbulence in the process leading to further homogenization mechanism. Various valves exist, such as counter-jet microchannel, radial diffuser annular valve, and axial flow through orifice valve. Different designs were tried out to improve the process of homogenization. Jet to the wall (Z chamber) was the modification done to reduce the increased shear on the wall and the rise in the temperature during the process. Y chamber was the modification for obtaining better and uniform droplet size distribution. Also, it has the added advantage of increasing the life of the chamber, as the high speed of the flow of the sample will lead to the increased collision of the particles. Some valves (counter-jet) showed the disadvantage of being unable to control the pressure. Hence amend ment done to control pressure was the development of a radial diffuser annular valve (Martínez-Monteagudo et al., 2016). However, increased particle collision with the walls of the chamber generates metal particle residues, lowering the emulsification efficiency. This is prevented by using diamond material to prevent the formation of metallic particles.
B. *Homogenization pressure* Increasing the pressure will increase the eddies in the process, causing increased interaction between the particles, thus resulting in reduced particle size. Laplace's pressure governs this correlation between the particle size and the pressure.

$$P(\text{inside}) - P(\text{outside}) = \Delta P = \gamma \frac{2}{a}$$

Where a is the radius of the particle and γ is surface tension. Pressure greater than Laplace's pressure will lead to disruption of the droplets(Yadav & Kale, 2020).
C. *Number of cycles* Increasing the number of cycles will subject the fluid to undergo the process of homogenization all over again and thus further reduce the size of the droplet.
D. *Temperature* Temperature of the process is determined based on the materials that are to be exposed to the process of high-pressure homogenization. For materials that have a low melting point or are thermolabile, cold homogenization is preferred. In contrast, hot homogenization is employed for materials where the lipid content is high, or the drug is to be entrapped into lipids. Thus, high-

pressure homogenization can be classified into hot and cold homogenization based on temperature.

The hot homogenization technique is a process which is carried out at a temperature above the melting point of the lipid and lipophilic drug present in the sample. The drug loading is carried out by melting the lipid and mixing it with the drug. This mixture is added to the aqueous phase under high pressure under constant temperature. The obtained mixture is nanoemulsion which, upon cooling down to room temperature, forms nanoparticles.

This system is thus suitable for manufacturing the particles that have high lipid concentrations and is considered a greener system as the process is free of organic solvent. Since this process comprises heating and cooling as the steps, it is comparatively less time-consuming, leading to lowered overall production time. The absence of any factor other than the rate of cooling and heating will affect the particle formation and vary the narrow particle size distribution obtained. Due to lower factors that are needed to be controlled in the whole process, the technique is easy to scale up. Just like every technique, even this technique has some limitations. Since the important parameter is heating the lipid to melt and dispersing the drug in the molten lipid, thermolabile drugs and low melting point lipids cannot be used for manufacturing the particles by this process. A significant amount of drug loss is observed as the drug is distributed into the aqueous phase to a greater extent. There is also a possibility of complexity during crystallization of nanoemulsions. Since it requires heating, it is an energy extensive process.

Cold homogenization is similar to the hot homogenization process, the only difference being not using heating as the parameter. The melted lipid drug mixture can cool and solidify to obtain microparticles. These are then subjected to high-speed stirring in cold aqueous surfactant, followed by the homogenization process. This process takes place at a temperature maintained either at room temperature or temperature below it. This leads to the conversion of microparticles to nanoparticles. Since this process does not require heating as a process parameter, thermolabile drugs, and materials can be used for manufacturing of nanoparticles. However, nanoparticles obtained via this method were found to depict broader size distribution (Xu et al., 2022).

2. *Microfluidic Technique*

Microfluidics is the flow-focusing method governed by rapid mixing and controlled diffusion (Ahn et al., 2018). It is the latest emerging technology that is used to prepare nanoparticles at a large scale. It is a process that mixes fluids at high pressure in a narrow environment (Chiesa et al., n.d.). For example, one of the inlets contains a lipid-stabilizing component in its aqueous phase, whereas the other inlet contains polymer and drug with the organic phase. At the junction of the two inlet channels, the nanoprecipitation process starts. These form polymer nanoparticles which are coated by lipid/ stabilizer. The particles are formed within milliseconds due to rapid mixing. The design of the microfluidic device can be as per the manufacturer's needs.

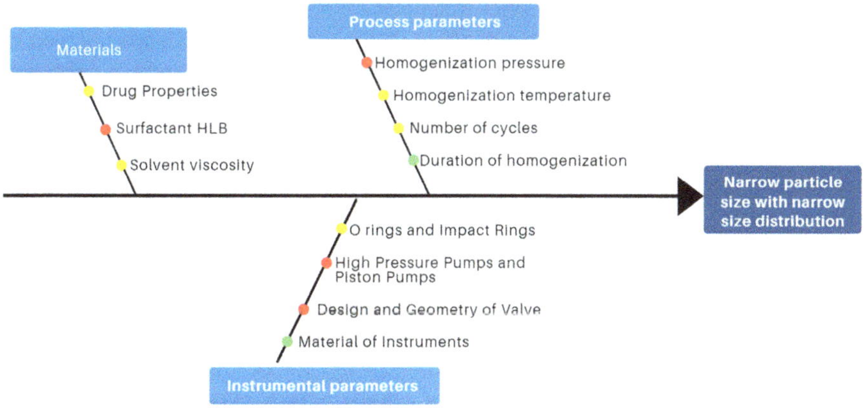

Fig. 12.6 Ishikawa diagram for High-pressure homogenization technique

The process parameters that will affect the mixing efficiency include (Herdiana et al., 2022):

A. *Operational characteristics* Reactant flow rate: It can be used for proper mixing. Pulsed micro mixing provides higher efficiency of mixing. The process consists of two components: a drive source (pump) and a Y-type micromixer. The higher total flow rate offers a lesser residence time of the drug molecule (or nuclei) in the chip that prevents agglomeration and subsequent clogging of the chip. Lababidi and co-workers observed similar effects of the reduction in the mean diameter of PLGA poly(lactic-co-glycolic acid) nanoparticles from 150 to 70 nm on increasing the flow rate of the external phase from 10 to 100 μL/min (Lababidi et al., 2019). Mixing time: It is another parameter that directly affects the dimensions of the precipitated particles. Short mixing time ensures homogenous mixing and thus leads to uniform particle size formation. It is, however, not possible for bulk production as uniform mixing at a larger scale needs a longer duration of time. The time can be kept short by modifying the design of the microreactors.
B. *Design characteristics* Type of microreactors: There are two types of microreactors: chip-based microfluidic devices and capillary-based microdevices. Capillary-based microdevices are further classified into continuous laminar flow microreactors, Segmented flow microreactors, and droplet-based microreactors (which consists of T-junction, Y-junction, flow focusing configuration, terrace-like device). These different microreactors will provide different efficiency of mixing. Even though capillary-based microdevice provides efficient mixing, it is not suitable for reactions that involve many factors that need to be controlled and thus making the reaction complicated. Thus, to solve this issue, chip-based microreactors were formed.

C. *Other factors* include channel diameter and the material with that the microreactors are made (this will affect the choice of solvent that is allowed to pass through the channels without causing corrosion or any other damage to the channel).

The advantages of using microfluidic systems include their ability to provide particular particle size, lesser variability among the batches, ability to be scaled up, narrow size distribution, cost-effective (especially for capillary-based microreactors), compatibility with various types of solvents, higher yield in comparison to smaller inlet volume, ability to combine synthesis, purification, analysis all in one "lab-on-chip" design, and ability to operate even at higher temperature and pressure conditions (Hao et al., 2018). Xu and co-workers manufactured PLGA nanoparticles loaded with either doxorubicin or tamoxifen. The optimized particle size was by altering the flow rate ratio and polymer concentration (Xu et al., 2017).

3. *Ultrasonication*

It is a technique generally used for size reduction. When low pressure is applied to liquid, there is a generation of high-intensity ultrasonic waves. These further create small vacuum bubbles, which expand to a certain extent and explode after that leading to the breakdown of the particles. This process is also termed cavitation. It is a top-down approach. The size of the nanoparticle depends upon the frequency of the sound waves and the intensity of ultrasonication. The advantage of using this technique is that it helps create lipid nanoparticles without any use of organic solvents. The time required to obtain nanoparticles is comparatively very less, and the technique is simple to use and efficient. The disadvantage of this technique is the broader particle size distribution obtained.

There is a possibility of metal impurities being generated from the sonication probe; thus, there is a need to add an additional step of filtration. The final product also faces some physical instability, such as the growth of particles when kept for storage. There are two types through which sonication can be done: probe sonication and bath sonication. At the same time, probe sonication makes use of a probe which consists of a replaceable tip through which the sound waves are transferred, whereas, bath sonication makes use of a container consisting of water which transmits the sound waves generated from the piezoelectric cell. A bath sonicator is preferred for dilute lipid dispersions, which are present in large volumes. The different factors that affect the size of the nanoparticles formed are the composition and concentration of lipids, the duration for which the process is operated, the power at which the sonicator is operated, the temperature used for sonication, etc.

12.2.1.3 Thermal Techniques

1. *Thermal Decomposition*

Thermal decomposition, also known as thermolysis, involves the chemical decomposition of bulk material via high temperatures to produce nanoparticles (Cele, 2020). Among the conventional methods, thermal decomposition is a convenient, fast-developing, and economical method (Odularu, 2018). Metal-ion complexes such as iron-oleate (Ullrich et al., 2019), copper oxalate (Salavati-Niasari et al., 2008a), [Bis(salicylidene) cobalt (II)] (Salavati-Niasari et al., 2008b), and Cu and Co tartrate complexes (Palacios-Hernández et al., 2012) are used as precursors for the synthesis of respective metal nanoparticles. The factors governing the uniform particle size distribution include duration of synthesis, temperature, concentrations of the reactants, stabilizers, capping agents (surfactants), and types of surfactants (Kino et al., 2008).

Aladin Ullrich and co-workers examined the effect of surfactant (oleic acid and sodium oleate) and heating rate on the particle size of iron oxide nanoparticles synthesized by the thermal decomposition method. A lower heating rate (i.e., 1 °C/min) produced larger particle size than a higher heating rate (i.e., 30 °C/min). The absence of surfactants produced spherical nanoparticles, whereas the anisotropic growth rate in the presence of surfactants produced cubical nanoparticles (Ullrich et al., 2019).

12.2.2 Bottom-Up Approach

Using non-polar solvents during chemical reduction synthesis further leads to environmental hazards. The high temperatures and pressures are frequent requirements during the "bottom-up", which add up the energy consumption and production costs. In terms of biomedical applications, the use of biocompatible chemicals is preferred to minimize the toxic effect and provide safe usage (Gajera et al., 2019).

12.2.2.1 Solvent Evaporation Techniques

1. *Spray Drying*

The spray drying technique not only deals with the rapid evaporation of solvents but can also be used for ultrafine powder formation. The selection of dryers would be based on the nature of starting material and end product. There are two types of spray dryers based on the direction of air flow and liquid atomization – concurrent flow (similar direction) and counter-current flow (different direction). The concurrent flow is the method of choice for drying heat-sensitive materials (Since the finished product is located in the region with the lowest temperature). However, in a counter-current flow, the dry product comes into contact with the warmest air, increasing thermal efficiency.

Process parameters that need to be considered while using the spray drying process are

A. *Inlet temperature of the drying gas* The temperature in the conventional dryer is set at 220 °C. A high inlet temperature guarantees a high rate of solvent evaporation, which affects the production of particles and the stability of the finished product. The higher temperature causes faster agglomeration of particles in comparison to low temperature. The surface of such particles may also be rough. Additionally, the kinetics of nucleation is controlled by the solvent evaporation rate, which is crucial for producing amorphous structures. The clogging and degradation of thermosensitive material are the main issues that might arise due to higher inlet temperatures. However, it should be highlighted that the droplet-to-particle transition takes milliseconds, limiting the chance of compound thermal degradation.

B. *Viscosity of the feed* The viscosity ideally chosen for conventional spray drying is below 300 mPas, and the solid content should fall below 30% since more energy is required for the atomization of the higher viscosity molecules. Along with this, the grade and concentration of polymeric material also play an important role. Sometimes, surfactants have also been used to reduce the interfacial tension and smaller droplet formation. The velocity is adjusted so that it should give narrow spray patterns.

C. *Other factors:* Other factors include feed type and composition, surface tension, sample volume, circulation pump rate, inlet temperature, chamber length, drying gas type (air, N_2/CO_2), drying gas humidity, aspirator speed, spray mesh size (Salama, 2020). Carmona et al. optimized the parameters like solid content, feed flow, temperature, atomization air flow, etc., for the spray drying process for carotenoids. The observation for the study could be summarized as increase in solid content increases the encapsulation efficiency and viscosity thus lowering the particle size (Carmona et al., 2018).

2. *Supercritical Fluid Technique*

A supercritical fluid is a substance that, at its critical temperature and pressure, exhibits both liquid and gaseous properties. Currently available conventional technique such as spray drying, freeze drying, milling has numerous drawback that leads to excessive use of solvent, structural changes, lesser drug entrapment, broad particle size distribution, thermal and mechanical degradation and, further formulation instability. Some conventional techniques can be used along with supercritical fluid technology, such as the conventional spray drying method combined with continuous supercritical CO_2 that provides accurate control of the final size and solid-state form of nanoparticles (Long et al., 2019). However, it eliminates these challenges, and the use of supercritical fluids (SCFs) affords potential for nanoparticle formation (Jaradat et al., 2021). Carbon dioxide is the most widely used supercritical fluid (31 °C and 73 bar) and can be used in three ways in nanoparticle processing (1) as a solvent to dissolve nonpolar solutes, (2) as an antisolvent via atomization process to precipitate drugs, and (3) as an adjuvant to promote higher liquid breakdown. Supercritical CO_2 is preferred over other hydrocarbons and fluorinated hydrocarbon fluids due to their concern related with reactivity, toxicity, flammability, and environmental hazards. They have several characteristics that are intermediate between those of gases and liquids, such as viscosities (0.01–0.03 mPa

Table 12.3 Summary of different drug nanoparticles using supercritical fluid CO_2 technology

Drug	Solvent/antisolvent/ additive	Average particle size	References
Sertraline hydrochloride	Solvent	185 ± 9 nm	Sodeifian et al. (2022)
	Antisolvent	102 ± 11 nm	
Capecitabine	Antisolvent	243.3 nm	Amani et al. (2021)
Amiodarone hydrochloride	Solvent	<50 nm	Sodeifian and Sajadian (2019)
Doxorubicin hydrochloride	Cosolvent	<250 nm	Peng et al. (2019)

and 0.4–1.1 mPa, respectively) and diffusivities (1–20.0 mm^2/s and 0.001–0.01 mm^2/s, respectively) that represent unique tuneable property useful for transport and rapid mixing. Their other properties, such as low surface tension values, low dielectric constant, excellent heat transfer and high compressibility, anti-oxidation effect towards the free radical reactions, and their aprotic nature, seem to be useful in different processes.

Additionally, they are safer alternatives to other organic solvents and are more affordable, reusable, recyclable, and environment friendly. It also has a few limitations that cannot be ignored. It cannot dissolve and extract polar substances, and non-polar substances that can be extracted require high power consumption. Also, the setup installation is complex and very expensive (al Khawli et al., 2019). Following is the recent literature applicable to the supercritical fluid CO_2 for the fabrication of nanoparticles (Table 12.3):

Process parameters that need to be considered during the preparation of nanoparticles are temperature, pressure, supersaturation degree, flow rate, residence time, nozzle, and expansion design chamber.

Supercritical fluid methods to date are not a scalable approach in the pharmaceutical industry. Nevertheless, there has been an upsurge of industrial interest in such upcoming technologies (Long et al., 2019). It is a scalable technique. However, it has the disadvantage of forming nanoparticles in batch size. New advancements overcome this in the technique, such as the rapid expansion of supercritical solutions, etc. The following cGMP is a must for any equipment that is used for the production of nanoparticles at a large scale. Additional safety requirements need to be added to the system as there are chances of vessel explosion due to equipment operation under high pressure. The setup operation needs to comply with the guides set out by the regulatory agencies in the respective countries (Padrela et al., 2018).

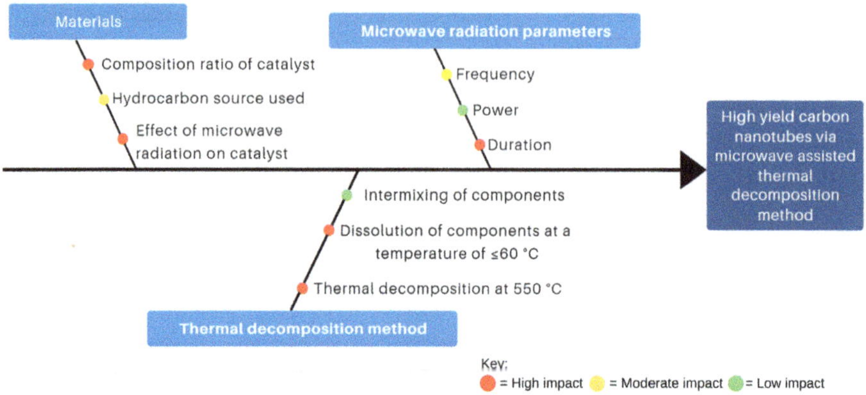

Fig. 12.7 Ishikawa diagram for chemical vapour deposition technique

12.2.2.2 Thermal Techniques

1. *Chemical Vapour Deposition*

The effect of microwave modification was also observed on the catalyst characteristics such as particle size distribution, average pore size, and specific pore volume (Fig. 12.7).

Carbon nanomaterials such as carbon nanotubes, carbon dots, etc. have been a fascinating system in the nano-drug delivery due to their exhibition of excellent chemical and physical properties, viz. high tensile strength, ultra-lightweight, exalted chemical, and thermal stability along with distinct electronic structures. Carbon nanotubes (CNT) have received more interest from a scientist in the R&D sector. CNT serves its purpose as a drug, providing targeted drug delivery to organs, a highly sensitive biosensor for diagnostic imaging and analysis (Anzar et al., 2020). Elena et al. have developed a novel method in addition to economic feasibility to synthesize CNT by modifying the Ni-MgO catalyst by microwave radiation (0.8 kW and 2.45 GHz). The catalyst characteristics (viz., specific surface area, catalytic activity, etc.) were enhanced by short-term exposure (30 s) to microwave radiation. The specific surface area of the Ni-MgO catalyst was increased from 5.2 to 9.1 m^2/g due to the use of short-term microwave radiation exposure. The effect of microwave modification was also observed on the catalyst characteristics such as particle size distribution, average pore size, and specific pore volume. The CNT yield has increased by 40–45% by using the modified Ni-MgO catalyst in CNT synthesis by the chemical vapor deposition method. The utilization of microwave radiation upsurges energy costs, though the use of microwaves is economically defensible and advantageous for large-scale production (Burakova et al., 2018). The process has the disadvantage of by-product formation, such as carbides, graphitic flakes, and CNT, leading to irreproducible yields. Conventional approaches like arc discharge have been engaged due to the economical production of carbon nanoparticles.

Jyongsik Jang and co-workers developed a method for multigram scale production of Iron encapsulated multi-walled carbon nanoparticles (MCNP) using micelle templating followed by a carbonization process (850 °C). The micelles were formed using dodecyl trimethyl ammonium bromide (DTAB) and decyl alcohol as a co-surfactant. The monomers diffusion from micelles was retarded by the co-surfactant, thereby providing size uniformity of the nanoparticles. The MCNP particle size depends on process parameters such as the micellar composition and the temperature of the reaction. The yield of polypyrrole nanoparticles (PPy) was >95%; these were used as a precursor for the formation of MCNP, which had a char yield of 50% (Jang & Yoon, 2005).

The chemical vapor desorption (CVD) technique is more advantageous than arc discharge and laser ablation according to cost scalability and operation temperature. Nevertheless, the operation temperature of CVD is very high (approximately 1000 ° C) in its synthesizing progress. This can be a problem while preparing particles containing materials that undergo degradation at such high temperatures, leading to impurities in the process. This can be solved using the plasma-enhanced CVD (PECVD) method, which operates at a much lower temperature and might lead to a faster growth rate. The mechanism is the rich and active electromagnetic environment produced by plasma consisting of free radicals, electrons, excited molecules, and other active particles.

The pure metals act as catalysts, including a substrate, and are placed on the substrate holder in the reaction chamber. Hydrocarbon gases frequently used as starting material the precursor for the synthesis of carbon nanoparticles are in the chamber before plasma. The plasma cavity is positioned in an attempt to cover the entire substrate. Turning on the plasma leads to (1) generation of a strong electromagnetic field which ionizes the hydrocarbon gas to form the reactive radicals and ions and (2) heating of the catalyst surface to accelerate the reaction. Due to the electromagnetic field, certain ions could be accelerated and bombard the substrate surface to create nucleation sites for the growth of carbon nanoparticles (Yeh et al., 2019). Factors such as reactor design, carbon source, catalyst category, the preparation method of catalysts, additive to a carbon source, precursor reduction temperature, catalyst additive, pyrolysis temperature, and reaction time have an impact on the growth of carbon nanomaterials (Qi et al., 2010). Figure 12.7 briefly summarizes the process parameters for chemical vapour deposition technique.

12.2.2.3 Chemical Techniques

The chemical method includes processes such as the chemical reduction method, photochemical method, and electrochemical method. Chemical method synthesis includes reducing the salt form to its elemental form. It is a reliable method as it provides the desired shape and size and can produce colloidal nanoparticles in organic solvents and water. It is usually carried out in the presence of three components: a salt precursor, a stabilizer, and a reducing agent. The process parameters that can affect the formation of nanoparticles are the type and ratio of precursor, reducing agents, temperature, and pH of the solution. Reduction of silver salt can also do with the help of light sources such as UV, sunlight, and laser light. Mostly, UV light is used for this process. It has the special advantage of producing in situ nanoparticles in the illumination region, which are highly dispersible. It provides the benefits of the ability to carry out experiments at room conditions and without the use of any harmful or strong reducing agent. If the light source is changed to microwave radiation, the type of synthesis is known as microwave-assisted synthesis. Generally, water and alcohol are considered to be the ideal media for the conduction of the experiment. This method has the advantage of being energy efficient, time-saving production, convenient in use, and can use at a large scale. Generally, water and alcohol are considered to be the ideal media for the conduction of the experiment. This method has the advantage of being energy efficient, time saving production, convenience in use, and has the ability to use at large scale. This technique is considered to be green in nature as it does not make use of any harmful solvents or any other techniques that will be considered harmful to the nature. It is not only green, but also simple to use, economical, and is environment friendly.

1. *Reduction*

Biological method synthesis makes us of microorganisms such as bacteria, fungi, algae, and plant-mediated synthesis. Plant mediates the reduction of the silver with the help of the presence of functional groups in the plant extract. In short, the plant extract acts as the reducing agent for the reduction of silver. The advantage of synthesizing silver nanoparticles by the biological method is that the reaction can be done at normal conditions of temperature and pressure. Whether it is bacteria-mediated synthesis, fungi-mediated, or algae-mediated, the presence of different organic matter in each of the organisms acts as reducing agents which help in the synthesis of silver nanoparticles. In the case of bacteria and fungi, synthesis can occur either intracellularly or extracellularly. The preferred option is extracellular preparation as the extraction of formed nanoparticles is easier from the outside. Pathogenicity is mostly associated with preparation in fungi and hence is avoided. Nanoparticles synthesized using plants, and algae can be of varying sizes and shapes. The process parameters that affect the synthesis of nanoparticles are temperature, pH, and concentration of plant extract and precursors (Xu et al., 2020) .

In 1951, Turkevich developed a method for the synthesis of the AuNP by the use of Sodium citrate as a reducing and stabilizing agent (Turkevich et al., 1951). The critical process parameters to be noted for Turkevich synthesis are pH, the order of reagent addition, and the latent heat of the reaction. Although this method is uncomplicated and reproducible for spherical AuNP with 10–30 nm size, the size distribution increases for AuNPs above 30 nm with deviation from spherical nature and low yield (Dong et al., 2020).

T. W. Nilsson et al. developed two types of gold nanoparticle at large scale using tannic acid (TA) and trisodium citrate (TSC) in a stirred tank reactor to yield nanoparticles (TA-AuNPs and TSC-AuNPs) of a size and shape distribution that was comparable to those produced on the laboratory-scale. TA and TSC played acts as a reducing and stabilizing the AuNPs. Unwanted gold deposition on the surface of the tank reactor was observed on a large scale (Nilsson & Johnston, n.d.).

The critical parameters to be controlled for the synthesis include the precursor solutions pH, mode of contacting, dynamics of stabilizer adsorption, and addition rate of chloroauric acid solution to tannic acid and trisodium citrate (Sivaraman et al., 2010).

Brust et al. synthesized narrow-size distribution gold nanoparticles by reduction of HAuCl4 with NaBH4 in the presence of alkanethiols (Brust et al., 1995). However, the lack of scalability approaches for the burst method could be due to economic aspects.

Lakshminarayana and the group developed a convenient and scalable production method for highly monodisperse gold nanoparticles at room temperature by mixing chloroauric acid (HAuCl4) and oleylamine. The reduction, as well as the capping of AuNPs, was carried out by oleylamine and particle size of 13nm was obtained. Molar ratio variation of HAuCl4 and oleylamine aids in the control of particle size. Replacing oleylamine as a capping agent with other groups, such as thiols, is possible for functionalization (Nilsson & Johnston, n.d.).

Robust and scalable method for preparation of AuNPs was developed by Tomáš Lednický et al. using solid-state dewetting on porous anodic alumina (PAA) and epoxy as templates. Selective chemical etching process was used for synthesis of templates. The method provided advantages such as (1) controlled particle size aiding the modulation of plasmonic absorption peak according to the applications; (2) large-scale fabrication as the lateral expansion of templates is possible, providing several cm^2 surface area for AuNPs synthesis; and (3) robustness due to fixation on epoxy template and avoiding exposure to the fluidic environment (Lednický & Bonyár, 2020).

12.2.2.4 Microemulsion Technique

A microemulsion is nothing but an emulsion with a globule size of microns. An emulsion is a dispersed system of two immiscible solvents mixed in the presence of an emulsifier. For converting the emulsion to microemulsion and forming a stable system, a component known as cosurfactant is added along with the existing emulsifier/surfactant. Other components are the same as emulsion, oil phase, and aqueous phase along with surfactant (Fig. 12.8).

The advantage of this manufacturing technique is that it does not need any energy input to prepare nanoparticles. Also, since there is no heating involved in this process, thermolabile drug and materials can be used for the preparation of nanoparticles. The process used is limited for use due to the necessity of the incorporation of high concentrations of cosurfactant and surfactant. Also, the solubility of the drug in either of the phase is needed to some extent to allow the formation of microemulsion and further formation of nanoparticles. Removal of excess water becomes a task from the lipid nanoparticles. Since solubility comes into the picture while the formation of nanoparticles, the pH of the medium is extremely significant and needs to be maintained, the temperature can also affect the stability of the microemulsion and thus affect the process of manufacturing nanoparticles.

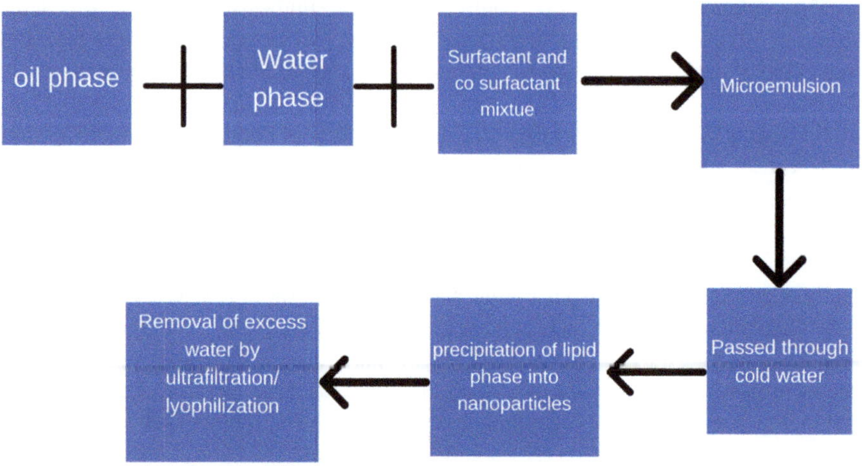

Fig. 12.8 Process of forming nanoparticles by microemulsion method

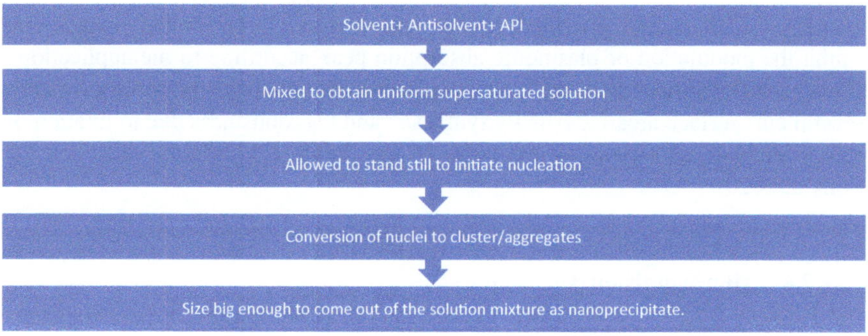

Fig. 12.9 Flow chart indicating a process of synthesis of nanoparticles by nanoprecipitation technique

12.2.2.5 Nanoprecipitation

Nanoprecipitation is a bottom-up approach which converts molecules into clusters and finally into nanoparticles. The process followed is as follows (Fig. 12.9):

The most crucial factor for obtaining nanoparticles is by controlling the nucleation process in the supersaturated solution. Phase separation can be obtained with the help of different factors such as pH, temperature, and antisolvent to modify the solubility of the particles. The use of antisolvent is considered a phase separation inducer in this method. The different process parameters that can affect the formation of nanoparticles are as follows (Hamdallah et al., 2020) (Fig. 12.10):

A. *Operational Characteristics* Degree of supersaturation The time spent by the particles in the nucleation phase will influence the size of the particles. A low degree of supersaturation will ultimately lead to large poly-dispersed particles as

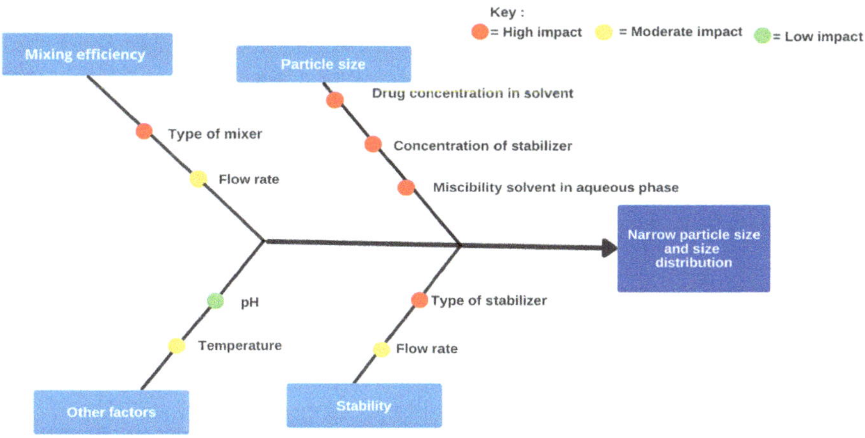

Fig. 12.10 Fishbone diagram summarizing the range of factors that can affect the formation of nanoparticles by the nanoprecipitation method. Colour dot indicates the level of risk assessment

compared to a higher degree of saturation. Thus, the optimum degree of supersaturation is preferred to achieve the expected particle size (Tao et al., 2019). Furthermore, it is optimized by altering the solvent or antisolvent phase, as demonstrated by Kakran et al., for quercetin nanoparticles by the antisolvent nanoprecipitation method. Thus, this results in the fabrication of polydisperse nanoparticles due to the increased antisolvent concentration leading to an enhanced Ostwald ripening effect in the dissolved drug molecule in the solvent (Kakran et al., 2012).

Miscibility of solvent in aqueous phase: Miscibility of solvent and aqueous phase and uniform mixing will contribute to the particle size. The highly miscible solvent will form smaller nanoparticles as compared to the solvent with lower miscibility. Tao et al. mentioned that for flash nanoprecipitation, it is important to control the rapid mixing of solution and antisolvent to achieve desirable supersaturation and for the growth of molecules by coagulation and condensation (Tao et al., 2019). Proper mixing will reduce the probability of aggregation as well as the possibility of Ostwald ripening (Thorat & Dalvi, 2012). Increasing the solute concentration in the solvent phase will increase the solute concentration in the binary mixture, thus increasing the degree of supersaturation and hence, producing small nanoparticles. Increasing the ratio of solvent and antisolvent will decrease the degree of supersaturation, thus increasing the solvent content in the binary mixture leading to stimulation of Ostwald ripening after particle precipitation (Gajera et al., 2019).

Type of stabilizer: Different stabilizers are used, some being surfactant (cationic, anionic, non-ionic) while some being polymers (van Eerdenbrugh et al., 2009). These stabilizers absorbed the formed particles, thus decreasing the surface free energy. This leads to a decrease in the tendency of those particles to interact further with each other and form aggregates. This is also prevented by stabilizing the nanoparticles formed either sterically or statically(Verma et al.,

2011). Stabilizer concentration is an essential factor as if the concentration of the surfactant exceeds the CMC concentration, Ostwald ripening is promoted, thus increasing the size of the particles, whereas if the concentration of the polymers used as a stabilizer increases above the critical flocculation concentration, polymers start to flocculate and thus entangle amongst each other leading to the aggregate formation (Thorat & Dalvi, 2012). However, this might not be followed by all the polymers or in all cases. Satranidazole, a high-dose poorly water-soluble drug optimized using nanoprecipitation technique by Dhat et al., in which PVA and Poloxamer were used as non-ionic stabilizers, results in increased diffusion of the drug into the external phase and a decrease in the particle size(Dhat et al., 2017).

Following are the process parameters that will control the particle size, and it is thus essential to consider them while scaling up the process. This technique is not used in industry as many factors need to be considered during the preparation of the particles, which can create many technical problems during the scale-up process. The particle size obtained is not uniform as multiple factors control the particle size. Controlling other factors, such as the temperature, pH, and miscibility of the solvent with the antisolvent phase and drug/excipient selection, becomes a task. Other than the factors mentioned above, the factors enlisted in the fishbone diagram below also affect particle size.

12.2.2.6 Biological Techniques

1. *Bioreduction*

Biological method synthesis makes use of microorganisms such as bacteria, fungi, algae, and plant-mediated synthesis. Microorganisms are capable of bioreduction with the aid of intracellular and extracellular bioreduction agents like NADH-reductase, phytochelatins, melanin, redox mediators, membrane proteins, cytochromes, and inorganic redox compounds. The biosynthesis of AuNP opens new doors for green method synthesis of AuNP. The controlled particle size, shape, and distribution of AuNP are due to specific interactions of Au with the bioreduction agents at the lab-scale level (Das et al., 2012b). To minimize the aggregation of AuNP, the extracellular proteins are conjugated to AuNPs by microorganisms. The process is known as bio-capping.

The challenges faced during scale up are poor reproducibility, slow reaction time, inadequate characterization of the formed particles, uncontrolled particle size and shape, the wide particle size distribution and cumbersome standardization procedures. The optimization of bioreactor operative conditions like pH, biomass, or Au^{3+} ratio affects the yield of AuNPs. The designing of the bioreactor for large-scale purposes turns into a major challenge as it should provide thin layer immobilization of the microorganism to minimize diffusional limitation and reduce the time needed for biosynthesis. Further simple recovery processes are required for an economic biosynthesis of AuNP (Das et al., 2012a).

12.3 Conclusion

Multiple methods have been tried to produce nanoparticles for large-scale production. However, the scope for further improvement in the preparation and yield of prepared nanoparticles persists. Different approaches have been exhausted to prepare nanoparticles on a lab scale as well as a commercial scale because of the advantages these nanoparticles present with. However, looking at the growing research in the field of nanoparticles and its interest for use in clinical applications, there is a possibility of translation of some of the techniques being adopted by the industry for the preparation of nanoparticles. The major hurdle in translation from lab to commercial scale will have to face the detailed data that the regulatory agencies will require for approving the newer medicine (containing nanoparticles) into the market.

References

ABRAXANE® – Official Patient & Caregiver Website. (n.d.). Available at: https://www.abraxane.com/. Accessed 28 Jan 2023.

Ahn, J., et al. (2018). Microfluidics in nanoparticle drug delivery; From synthesis to pre-clinical screening. *Advanced Drug Delivery Reviews, 128*, 29–53. Available at: https://doi.org/10.1016/J.ADDR.2018.04.001

al Khawli, F., et al. (2019). Innovative green technologies of intensification for valorization of seafood and their by-products. *Marine Drugs*. Available at: https://doi.org/10.3390/md17120689

Amani, M., Saadati Ardestani, N., & Majd, N. Y. (2021). Utilization of supercritical CO_2 gas antisolvent (GAS) for production of Capecitabine nanoparticles as anti-cancer drug: Analysis and optimization of the process conditions. *Journal of CO_2 Utilization, 46*. Available at: https://doi.org/10.1016/J.JCOU.2021.101465/REFERENCES

AmBisome® | AmBisome (amphotericin B) liposome for injection. (n.d.). Available at: https://www.ambisome.com/. Accessed 28 Jan 2023.

Anzar, N., et al. (2020). Carbon nanotube – A review on synthesis, properties and plethora of applications in the field of biomedical science. *Sensors International, 1*, 100003. Available at: https://doi.org/10.1016/J.SINTL.2020.100003

Bitterlich, A., et al. (2014). Challenges in nanogrinding of active pharmaceutical ingredients. *Chemical Engineering & Technology, 37*(5), 840–846. Available at: https://doi.org/10.1002/CEAT.201300697

Brust, M., et al. (1995). Synthesis and reactions of functionalised gold nanoparticles. *Journal of the Chemical Society, Chemical Communications, 16*, 1655–1656. Available at: https://doi.org/10.1039/C39950001655

Burakova, E. A., et al. (2018). Novel and economic method of carbon nanotubes synthesis on a nickel magnesium oxide catalyst using microwave radiation. *Journal of Molecular Liquids, 253*, 340–346. Available at: https://doi.org/10.1016/J.MOLLIQ.2018.01.062

Carmona, P. A. O., et al. (2018). Effect of solids content and spray-drying operating conditions on the carotenoids microencapsulation from pressed palm fiber oil extracted with supercritical CO_2. *Food and Bioprocess Technology, 11*(9), 1703–1718. Available at: https://doi.org/10.1007/S11947-018-2132-3

Cele, T. (2020). Preparation of nanoparticles. *Engineered Nanomaterials – Health and Safety* [Preprint]. Available at: https://www.academia.edu/56977336/Preparation_of_Nanoparticles. Accessed 28 Jan 2023.

Chiesa, E., et al. (n.d.). The microfluidic technique and the manufacturing of polysaccharide nanoparticles. *Pharmaceutics*. Available at: https://doi.org/10.3390/pharmaceutics10040267

COPAXONE® | Medication to Treat Relapsing Multiple Sclerosis (RMS). (n.d.). Available at: https://www.copaxone.com/. Accessed 28 Jan 2023.

Das, S. K., Liang, J., et al. (2012a). Biomineralization mechanism of gold by zygomycete fungi rhizopous oryzae. *ACS Nano, 6*(7), 6165–6173. Available at: https://doi.org/10.1021/NN301502S/SUPPL_FILE/NN301502S_SI_001.PDF

Das, S. K., Dickinson, C., et al. (2012b). Synthesis, characterization and catalytic activity of gold nanoparticles biosynthesized with Rhizopus oryzae protein extract. *Green Chemistry, 14*(5), 1322–1334. Available at: https://doi.org/10.1039/C2GC16676C

Dhat, S., et al. (2017). Risk management and statistical multivariate analysis approach for design and optimization of satranidazole nanoparticles. *European Journal of Pharmaceutical Sciences, 96*, 273–283. Available at: https://doi.org/10.1016/J.EJPS.2016.09.035

Dong, J., et al. (2020). Synthesis of precision gold nanoparticles using Turkevich method. *Kona: Powder Science and Technology in Japan, 37*, 224–232. Available at: https://doi.org/10.14356/KONA.2020011

Ferinject (ferric carboxymaltose) – Summary of Product Characteristics (SmPC) – (emc). (n.d.). Available at: https://www.medicines.org.uk/emc/product/5910/smpc#gref. Accessed 28 Jan 2023.

Gajera, B. Y., Shah, D. A., & Dave, R. H. (2019). Development of an amorphous nanosuspension by sonoprecipitation-formulation and process optimization using design of experiment methodology. *International Journal of Pharmaceutics, 559*, 348–359. Available at: https://doi.org/10.1016/J.IJPHARM.2019.01.054

Galen acquires DaunoXome® (daunorubicin citrate liposome injection) for advanced HIV-associated Kaposi's sarcoma | Fierce Biotech. (n.d.). Available at: https://www.fiercebiotech.com/biotech/galen-acquires-daunoxome%C2%AE-daunorubicin-citrate-liposome-injection-for-advanced-hiv. Accessed 28 Jan 2023.

Hamdallah, S. I., et al. (2020). Microfluidics for pharmaceutical nanoparticle fabrication: The truth and the myth. *International Journal of Pharmaceutics, 584*. Available at: https://doi.org/10.1016/J.IJPHARM.2020.119408

Hao, N., Nie, Y., & Zhang, J. X. J. (2018). Microfluidic synthesis of functional inorganic micro−/nanoparticles and applications in biomedical engineering. *International Materials Reviews, 63*(8), 461–487. Available at: https://doi.org/10.1080/09506608.2018.1434452

Herdiana, Y., et al. (2022). Scale-up polymeric-based nanoparticles drug delivery systems: Development and challenges. *OpenNano (Elsevier Inc)*. Available at: https://doi.org/10.1016/j.onano.2022.100048

Howard, J. L., Cao, Q., & Browne, D. L. (2018). Mechanochemistry as an emerging tool for molecular synthesis: What can it offer? *Chemical Science (Royal Society of Chemistry), 9*(12), 3080–3094. Available at: https://doi.org/10.1039/c7sc05371a

Jang, J., & Yoon, H. (2005). Multigram-scale fabrication of monodisperse conducting polymer and magnetic carbon nanoparticles. *Small, 1*(12), 1195–1199. Available at: https://doi.org/10.1002/smll.200500237

Jaradat, E., et al. (2021). Microfluidics technology for the design and formulation of nanomedicines. *Nanomaterials (MDPI)*. Available at: https://doi.org/10.3390/nano11123440

Kakran, M., et al. (2012). Fabrication of quercetin nanoparticles by anti-solvent precipitation method for enhanced dissolution. *Powder Technology, 223*, 59–64. Available at: https://doi.org/10.1016/J.POWTEC.2011.08.021

Kino, T., et al. (2008). Synthesis of chalcopyrite nanoparticles via thermal decomposition of metal-thiolate. *Materials Transactions, 49*(3), 435–438. Available at: https://doi.org/10.2320/MATERTRANS.MBW200724

Kobayashi, M., et al. (2017). Effect of ball milling on the physicochemical properties of atorvastatin calcium sesquihydrate: The dissolution kinetic behaviours of milled amorphous solids. *Journal of Pharmacy and Pharmacology, 69*(1), 15–22. Available at: https://doi.org/10.1111/jphp.12636

Kumar, M., et al. (2020). Ball milling as a mechanochemical technology for fabrication of novel biochar nanomaterials. *Bioresource Technology, 312*. Available at: https://doi.org/10.1016/J.BIORTECH.2020.123613

Lababidi, N., et al. (2019). Microfluidics as tool to prepare size-tunable PLGA nanoparticles with high curcumin encapsulation for efficient mucus penetration. *Beilstein Journal of Nanotechnology, 10*(1), 2280–2293. Available at: https://doi.org/10.3762/BJNANO.10.220

Lednický, T., & Bonyár, A. (2020). Large scale fabrication of ordered gold nanoparticle-epoxy surface nanocomposites and their application as label-free plasmonic DNA biosensors. *ACS Applied Materials and Interfaces, 12*(4), 4804–4814. Available at: https://doi.org/10.1021/ACSAMI.9B20907/ASSET/IMAGES/LARGE/AM9B20907_0007.JPEG

Liu, X., & Meng, H. (2021). Consideration for the scale-up manufacture of nanotherapeutics – A critical step for technology transfer. *VIEW, 2*(5), 20200190. Available at: https://doi.org/10.1002/viw.20200190

Loh, Z. H., Samanta, A. K., & Sia Heng, P. W. (2014). Overview of milling techniques for improving the solubility of poorly water-soluble drugs. *Asian Journal of Pharmaceutical Sciences (Shenyang Pharmaceutical University)*, 255–274. Available at: https://doi.org/10.1016/j.ajps.2014.12.006

Long, B., Ryan, K. M., & Padrela, L. (2019). From batch to continuous – New opportunities for supercritical CO_2 technology in pharmaceutical manufacturing. *European Journal of Pharmaceutical Sciences, 137*. Available at: https://doi.org/10.1016/J.EJPS.2019.104971

Mariño-Gámez, A. E., et al. (2022). Influence of high energy ball milling on structural, microstructural and optical properties of TiO_2 nanoparticles. *Ceramics International, 48*(3), 3362–3367. Available at: https://doi.org/10.1016/J.CERAMINT.2021.10.111

Martínez-Monteagudo, S. I., Yan, B., & Balasubramaniam, V. M. (2016). Engineering process characterization of high-pressure homogenization – From laboratory to industrial scale. *Food Engineering Reviews, 9*(3), 143–169. Available at: https://doi.org/10.1007/S12393-016-9151-5

Nanoparticles from Mechanical Attrition. (n.d.). Available at: https://www.researchgate.net/publication/246917719_Nanoparticles_from_Mechanical_Attrition. Accessed 28 Jan 2023.

Nilsson, T. W., & Johnston, J. H. (n.d.). *Making nanoparticles mega: Scaling-up the synthesis of gold nanoparticle-wool fibre composites*. Available at: www.nsti.org. Accessed 28 Jan 2023.

Odularu, A. T. (2018). Metal nanoparticles: Thermal decomposition, biomedicinal applications to cancer treatment, and future perspectives. *Bioinorganic Chemistry and Applications*. Available at: https://doi.org/10.1155/2018/9354708

ONCASPAR – Use in Treating Acute Lymphoblastic Leukemia (ALL). (n.d.). Available at: https://www.oncaspar.com/. Accessed 28 Jan 2023.

P. R. Rittinger, "Lehrbuch der Aufbereitungskunde," Berlin, 1867 – References – Scientific Research Publishing. (n.d.). Available at: https://www.scirp.org/(S(vtj3fa45qm1ean45vvffcz55))/reference/referencespapers.aspx?referenceid=180611. Accessed 28 Jan 2023.

Padrela, L., et al. (2018). Supercritical carbon dioxide-based technologies for the production of drug nanoparticles/nanocrystals – A comprehensive review. *Advanced Drug Delivery Reviews, 131*, 22–78. Available at: https://doi.org/10.1016/J.ADDR.2018.07.010

Palacios-Hernández, T., et al. (2012). Synthesis of Cu and Co metal oxide nanoparticles from thermal decomposition of tartrate complexes. *Inorganica Chimica Acta, 392*, 277–282. Available at: https://doi.org/10.1016/J.ICA.2012.03.039

Peng, H. H., et al. (2019). Preparation of pH-responsive DOX-loaded chitosan nanoparticles using supercritical assisted atomization with an enhanced mixer. *International Journal of Pharmaceutics, 558*, 82–90. Available at: https://doi.org/10.1016/J.IJPHARM.2018.12.077

Prasad Yadav, T., Manohar Yadav, R., & Pratap Singh, D. (2012). Mechanical milling: A top down approach for the synthesis of nanomaterials and nanocomposites. *Nanoscience and Nanotechnology, 2*(3), 22–48. Available at: https://doi.org/10.5923/J.NN.20120203.01

Process Development & Scale-up – www.vbpgroup-ict.in. (n.d.). Available at: http://www.vbpgroup-ict.in/research/process-development-scale-up. Accessed 26 Jan 2023.

Qi, X., et al. (2010). Large-scale synthesis of carbon nanomaterials by catalytic chemical vapor deposition: A review of the effects of synthesis parameters and magnetic properties. *Materials, 3*(8), 4142–4174. Available at: https://doi.org/10.3390/MA3084142

RAPAMUNE® | Pfizer. (n.d.). Available at: https://www.pfizer.com/products/product-detail/rapamune. Accessed 28 Jan 2023.

Salama, A. H. (2020). Spray drying as an advantageous strategy for enhancing pharmaceuticals bioavailability. *Drug Delivery and Translational Research, 10*(1). Available at: https://doi.org/10.1007/S13346-019-00648-9

Salavati-Niasari, M., et al. (2008a). Preparation of cobalt nanoparticles from [bis(salicylidene)-cobalt(II)]–oleylamine complex by thermal decomposition. *Journal of Magnetism and Magnetic Materials, 320*(3–4), 575–578. Available at: https://doi.org/10.1016/J.JMMM.2007.07.020

Salavati-Niasari, M., Davar, F., & Mir, N. (2008b). Synthesis and characterization of metallic copper nanoparticles via thermal decomposition. *Polyhedron, 27*(17), 3514–3518. Available at: https://doi.org/10.1016/J.POLY.2008.08.020

Sandimmune (Cyclosporine): Uses, Dosage, Side Effects, Interactions, Warning. (n.d.). Available at: https://www.rxlist.com/sandimmune-drug.htm#warnings. Accessed 28 Jan 2023.

Shegokar, R., & Nakach, M. (2020). Large-scale manufacturing of nanoparticles – An industrial outlook. In *Drug delivery aspects: Volume 4: Expectations and realities of multifunctional drug delivery systems* (pp. 57–77). Elsevier. Available at: https://doi.org/10.1016/B978-0-12-821222-6.00004-X

Sherif El-Eskandarany, M., et al. (2021). Mechanical milling: A superior nanotechnological tool for fabrication of nanocrystalline and nanocomposite materials. *Nanomaterials, 11*(10), 2484. Available at: https://doi.org/10.3390/NANO11102484

Sivakumar, M., Tang, S. Y., & Tan, K. W. (2014). Cavitation technology – A greener processing technique for the generation of pharmaceutical nanoemulsions. In *Ultrasonics sonochemistry* (pp. 2069–2083). Elsevier B.V. Available at: https://doi.org/10.1016/j.ultsonch.2014.03.025

Sivaraman, S. K., Kumar, S., & Santhanam, V. (2010). Room-temperature synthesis of gold nanoparticles-size-control by slow addition. *Gold Bulletin, 43*. Available at: http://rsbweb.nih.gov/. Accessed 28 Jan 2023.

Soares, S., et al. (2018). Nanomedicine: Principles, properties, and regulatory issues. *Frontiers in Chemistry, 6*(AUG), 360. Available at: https://doi.org/10.3389/FCHEM.2018.00360/BIBTEX

Sodeifian, G., & Sajadian, S. A. (2019). Utilization of ultrasonic-assisted RESOLV (US-RESOLV) with polymeric stabilizers for production of amiodarone hydrochloride nanoparticles: Optimization of the process parameters. *Chemical Engineering Research and Design, 142*, 268–284. Available at: https://doi.org/10.1016/J.CHERD.2018.12.020

Sodeifian, G., Sajadian, S. A., & Derakhsheshpour, R. (2022). CO_2 utilization as a supercritical solvent and supercritical antisolvent in production of sertraline hydrochloride nanoparticles. *Journal of CO2 Utilization, 55*, 101799. Available at: https://doi.org/10.1016/J.JCOU.2021.101799

Subhan, M., Choudhury, K., & Neogi, N. (2021). Advances with molecular nanomaterials in industrial manufacturing applications. *Nano, 1*(2), 75–97. Available at: https://doi.org/10.3390/nanomanufacturing1020008

Tao, J., Chow, S. F., & Zheng, Y. (2019). Application of flash nanoprecipitation to fabricate poorly water-soluble drug nanoparticles. *Acta Pharmaceutica Sinica B, 9*(1), 4–18. Available at: https://doi.org/10.1016/J.APSB.2018.11.001

Thorat, A. A., & Dalvi, S. v. (2012). Liquid antisolvent precipitation and stabilization of nanoparticles of poorly water soluble drugs in aqueous suspensions: Recent developments and future perspective. *Chemical Engineering Journal, 181–182*, 1–34. Available at: https://www.academia.edu/34711942/Liquid_antisolvent_precipitation_and_stabilization_of_nanoparticles_of_poorly_water_soluble_drugs_in_aqueous_suspensions_Recent_developments_and_future_perspective. Accessed 28 Jan 2023

Tricor for Cholesterol: Uses, Dosage, Side Effects, Interactions, Warnings. (n.d.). Available at: https://www.rxlist.com/tricor-drug.htm#dosage. Accessed 28 Jan 2023.

Turkevich, J., Stevenson, P. C., & Hillier, J. (1951). A study of the nucleation and growth processes in the synthesis of colloidal gold. *Discussions of the Faraday Society, 11*(0), 55–75. Available at: https://doi.org/10.1039/DF9511100055

Ullrich, A., et al. (2019). Synthesis and high-resolution structural and chemical analysis of iron-manganese-oxide core-shell nanocubes. *Scientific Reports, 9*(1), 1–9. Available at: https://doi.org/10.1038/s41598-019-55397-z

van Eerdenbrugh, B., et al. (2009). A screening study of surface stabilization during the production of drug nanocrystals. *Journal of Pharmaceutical Sciences, 98*(6), 2091–2103. Available at: https://doi.org/10.1002/JPS.21563

Verma, S., et al. (2011). Physical stability of nanosuspensions: Investigation of the role of stabilizers on Ostwald ripening. *International Journal of Pharmaceutics, 406*(1–2), 145–152. Available at: https://doi.org/10.1016/J.IJPHARM.2010.12.027

Waje, S. B., et al. (2010). X-ray diffraction studies on crystallite size evolution of $CoFe_2O_4$ nanoparticles prepared using mechanical alloying and sintering. *Applied Surface Science, 256*(10), 3122–3127. Available at: https://doi.org/10.1016/J.APSUSC.2009.11.084

Wills, B. A., & Finch, J. A. (2016). Comminution. *Wills' Mineral Processing Technology*, 109–122. Available at: https://doi.org/10.1016/B978-0-08-097053-0.00005-4

XIPERE®, VISUDYNE® and RETISERT® for Retinal Condition Management: Bausch + Lomb. (n.d.). Available at: https://www.bausch.com/products/rx-pharmaceuticals/retinal-conditions/. Accessed 28 Jan 2023.

Xu, J., et al. (2017). Controllable microfluidic production of drug-loaded PLGA nanoparticles using partially water-miscible mixed solvent microdroplets as a precursor. *Scientific Reports, 7*(1), 1–12. Available at: https://doi.org/10.1038/s41598-017-05184-5

Xu, L., et al. (2020). Silver nanoparticles: Synthesis, medical applications and biosafety. *Theranostics, 10*(20), 8996–9031. Available at: https://doi.org/10.7150/THNO.45413

Xu, L., et al. (2022). Lipid nanoparticles for drug delivery. *Advanced NanoBiomed Research, 2*(2), 2100109. Available at: https://doi.org/10.1002/anbr.202100109

Yadav, K. S., & Kale, K. (2020). High pressure homogenizer in pharmaceuticals: Understanding its critical processing parameters and applications. *Journal of Pharmaceutical Innovation (Springer)*, 690–701. Available at: https://doi.org/10.1007/s12247-019-09413-4

Yeh, N. C., et al. (2019). Single-step growth of graphene and graphene-based nanostructures by plasma-enhanced chemical vapor deposition. *Nanotechnology, 30*(16). Available at: https://doi.org/10.1088/1361-6528/AAFDBF

Zolriasatein, A., et al. (2018). Comparative study of SPEX and planetary milling methods for the fabrication of complex metallic alloy nanoparticles. *Micro & Nano Letters, 13*(4), 448–451. Available at: https://doi.org/10.1049/MNL.2017.0608

Chapter 13
Manufacturing and Scale-Up of Biotechnology-Derived Products

Anil B. Jindal and Sagar S. Bachhav

Contents

Abstract In the last decade, there has been a notable increase in the approval of biotechnology-based pharmaceutical products by regulatory agencies for treating various diseases. Large-scale production of these products is crucial to meet the market demand. However, the complex nature of biological techniques presents challenges in scaling up these products obtained from cell culture techniques. This chapter delves into the typical unit operations involved in the manufacturing of biotechnology-based products and the scale-up considerations required to increase the batch size. It focuses explicitly on cell culture processes and does not include fermentation processes in its discussion.

Keywords Biotechnology · Bioreactor · Sterilization · Culture medium · Recombinant DNA technology · Purification

---

A. B. Jindal (✉)
Department of Pharmacy, Birla Institute of Technology and Science, Pilani, Jhunjhunu, Rajasthan, India
e-mail: anil.jindal@pilani.bits-pilani.ac.in

S. S. Bachhav
Clinical Pharmacology and Pharmacometrics, AbbVie, Inc., North Chicago, IL, USA

A. B. Jindal (ed.), *Pharmaceutical Process Engineering and Scale-up Principles*, AAPS Introductions in the Pharmaceutical Sciences 13,
https://doi.org/10.1007/978-3-031-31380-6_13

13.1 Introduction

Biotechnology-based pharmaceuticals are obtained from biological sources, such as living cells, through the use of advanced technologies. The advancements in recombinant DNA technology, cell culture techniques, and purification and characterization methods have led to a higher production of these therapeutic agents (Deb et al., 2019). The main advantage of these agents over traditional small molecules is their increased target specificity and fewer side effects. However, biotechnology-based therapeutic agents also present challenges, such as immunogenicity and injection site reactions, which are not seen in small molecular drugs. Additionally, the purification and characterization of these agents require advanced instrumental techniques, leading to higher production costs.

Furthermore, complex biological process demands skilled professional, and the low yield of the biological product further requires more advanced technology to meet the demands. Another major challenge involved in using biologics is ethical and regulatory concerns and significant scale-up challenges due to the complex nature of the manufacturing process, which ultimately delays the regulatory approval of the products (Gorsky, 2005). Biotechnology-based products have a significant market share, which includes products belonging to different categories, including antibiotics, hormones, enzymes, blood clotting factors, cytokines, monoclonal antibodies, and vaccines (Deb et al., 2019). The chapter mainly covers typical unit operations involved in the manufacturing of biotechnology-based products and scale-up considerations required to increase the batch size. The chapter mainly focuses on cell culture processes, and fermentation processes have not been included here.

13.2 Manufacturing of Biotechnology-Based Pharmaceutical Products

The manufacturing of biotechnology-based products is divided into two stages: upstream and downstream. The upstream process involves synthesizing the desired substance in a host cell using typical unit operations such as medium preparation, bioreactor operation, and filtration/centrifugation. The product obtained in the desired quantity in the upstream stage is then further processed by downstream processing, which includes isolation and purification using specialized techniques, such as filtration, centrifugation, and chromatography. The main goal of the entire process is to produce the biological product in the required quantities and purity. A thorough understanding of the biochemical process involved in synthesizing the biological product is essential for successful production. The following section provides an overview of the typical unit operations involved in the manufacturing of biological products.

13.2.1 Cell Culture Technology

Biotechnology-based products are obtained by allowing cells to grow in a culture medium in a bioreactor. The culture medium comprises various components that give the cells essential nutrients for growth. During this process, the cells consume nutrients from the culture medium and produce the desired product and waste materials. One of the most essential components of the culture medium is serum, which supports cell growth by providing essential substances. It usually adds up to 10% to the basal culture medium to ensure sufficient cell growth. The serum contains a variety of factors that contribute to cell growth, including proteins (albumin, lipoprotein, and transferrin), attachment factors, defense molecules, antibodies, complementary factors, and free radical scavengers. Different animal-derived serums, such as calf serum, fetal bovine serum, and horse serum, supplement the basal culture medium. In bioreactors, the cells remain suspended in the culture medium by agitation, which ensures uniform distribution of the nutrients for optimal cell growth (Spier, 2003).

While agitation is necessary for optimal cell growth, it can also produce mechanical shear stress on the cells, potentially causing damage. The agitation rate also has a significant effect on cell growth, metabolism, and cell surface receptor density (Cherry & Papoutsakis, 1986; Hu et al., 2011). Aerobic cells require molecular oxygen to produce energy. Still, due to its poor solubility in the liquid medium, specialized gas delivery devices and mild agitation are necessary to provide oxygen to the cells. The properties of the gas-liquid interface and mass transfer rate play a crucial role in oxygen delivery to the cells in the culture medium (Yasin et al., 2019). To avoid mechanical damage to the cells, the addition of surfactants and mild shear impellers may be employed (Chisti, 2000). The growth of cells is also affected by the physicochemical properties of the culture medium, including ionic strength, osmolarity, temperature, and pH. The osmolarity of the cell culture medium is typically adjusted to support optimal cell growth in the range of 270–330 mOsmol/kg. Both hyperosmotic and hypoosmotic culture mediums showed a decrease in cell growth and cell density. However, applying osmotic stress increases the production rate of biological substances. Temperature also plays a very important role in the production of biotechnology-based products from cell culture techniques. Generally, the temperature is controlled to 37 °C for optimum cell growth. However, temperature stress (deviation of the temperature from 37 °C) has shown a tremendous impact on the cell cycle and cell growth, product formation shear sensitivity of the cells. Additionally, pH control during production is also essential to enable optimum production. The generation of pH gradient could affect cell differentiation and cell growth, metabolism, and product formation (Nienow, 2006).

13.2.2 Bioreactor Operations

Bioreactors are vessels that hold cells and culture medium during the production of biotechnology-derived products. Typical bioreactor operations include cleaning, sterilization, charging, culture initiation, and harvesting. Cleaning of bioreactors is essential to ensure the purity and batch-to-batch reproducibility of biotechnology-based products (Bareither & Pollard, 2011). After completion of each batch, cleaning of bioreactors is accomplished using deionized water and 0.5 N sodium hydroxide for 1–4 h at 40–60 °C under high-speed stirring and mild airflow rate. The cleaning method validation is also essential to ensure the absence of batch-to-batch or product-to-product carryover (Spier, 2003). Standard sterilization methods including wet heat sterilization using autoclave, dry heat, hot air, filtration, and radiation are used to sterilize equipment parts, vessels, and culture medium (Jildeh et al., 2021). The bioreactor is charged with the liquid media components and base for pH control. Culture initiation involves transferring the cell inoculum and mixing. During the transfer of cell inoculum, there should be no physiological stress and the transfer process should be completed in a minimum time without generating any foam. During the initial mixing stage, just after inoculum transfer, the culture medium should be agitated at a minimum rate required to keep the cells suspended. Initial sampling is done for sterility testing, while a small sample of inoculum is retained to assess cell viability.

After the release of the product into the culture medium, it is isolated from the bioreactor, which is known as harvesting. Cell damage is a main concern during harvesting, which can be avoided by using low-shear transfer processes such as peristaltic pumps, pneumatic transfer, and gravity transfer. Additionally, the harvesting process is performed at low temperatures to prevent product degradation. Filtration and centrifugation are commonly used for clarification of the culture medium (Duong et al., 2019).

13.2.3 Filtration and Centrifugation

Centrifugation is used for clarification of the medium, while filtration is used to remove cells. Crossflow filtration is commonly used for cell separation, but it may result in cell lysis, which leads to a decrease in the yield and purity of the product (Clavijo Rivera et al., 2020). Centrifugation is a high-throughput, low-shear process preferred for large-scale processes. Filtration, on the other hand, is employed for small-scale batches as the high flow rate required at a large scale can increase shear stress on the cells, causing cell lysis. Tangential flow filtration is limited in separating animal cells due to cell damage and lysis. The main mechanisms for cell damage are laminar shear or deformation of the cells when passing through the pores of the filters. Several parameters should be optimized in the filtration process, including tangential flow rate, transmembrane pressure, permeate flux, pump speed, and type of flow (turbulent vs laminar) (Maiorella et al., 1991). Generally, during scale-up, the ratio between retentive flow to permeate flow should be 5:1 to avoid "dead-end filtration" when tangential flow filtration is used. Additionally, the ratio of filtration to process volume decreases as the batch size increases (Gorsky, 2005).

13.2.4 Chromatography

During the downstream processing, chromatography is used for the purification of the products. There are several chromatographic techniques available for the separation and purification, for example, size exclusive chromatography, ion exchange chromatography, and affinity chromatography. The primary advantage of the chromatographic technique is its high selectivity and sensitivity. The selection of an appropriate chromatographic technique depends upon the physicochemical properties of the substance which is to be isolated. For instance, if the product exhibit charge, in that case, ion exchange chromatography would be an ideal choice. In contrast, affinity chromatography is used to separate the substances that are hydrophobic in nature (Deb et al., 2019).

Size exclusion chromatography separates the substances based on the molecular size of the product. Porous beads are generally used as stationary phase selection of appropriate gel for the separation is essential (Ó'Fágáin et al., 2011). Ion exchange chromatography is generally used to separate biomolecular that exhibit charge. Resins with positively or negatively charged groups are used as a stationary phase, and the selection of the stationary phase depends upon the charge present in the biomolecules. Electrostatic interaction between the stationary phase and biomolecule is the basis for the separation (Fu et al., 2003). Affinity chromatography is another most frequently used chromatographic technique used in the purification of biomolecules in the biotechnology processes. Proteins are separated based on the interaction and affinity of the biomolecules with the stationary phase. Selecting an appropriate stationary phase is essential for the complete separation of the substance (Tanaka et al., 1993).

13.3 Scale-Up Considerations

Since biomolecule generation depends upon several parameters, including shear forces and physicochemical properties of the culture medium, the scale-up of biotechnology-based products is complex compared to the conventional formulation. The following section elaborates on various considerations involved in the production of these products.

13.3.1 Cell Culture Process Scale-Up

The primary challenge in scaling up the cell culture process is providing oxygen to the cells at the desired concentration, which becomes difficult at a large scale due to the poor solubility of oxygen in the culture medium. Additionally, achieving a homogenous distribution of nutrients in the culture medium requires turbulent mixing at a large scale, which can produce shear stress and damage cells mechanically. To avoid the development of a substantial gradient in bioreactor conditions and excessive shear stress on the cells, the following general scale-up rules are typically adopted:

(a) Addition of surface-active agents is essential during large-scale production to avoid cell damage.
(b) Typical impellers, including Rushton, pitched blade, or marine impeller, should be used for mixing in large-scale bioreactors.
(c) Direct gas sparging to the culture medium should be preferred to provide oxygen in substantial concentration to the cells.

The first-principle approach is also used for the scale-up of the cell culture process (Spier, 2003). To use the first-principle approach, the following points should be considered.

(a) A turbulent flow pattern exists in the bioreactors, which is far from typical random.
(b) Complete understanding of hydrodynamic forces acting on animal cells is always essential for all locations.

The main challenge at large-scale production is cell damage due to foam formation or bubble rupture. To this end, substances other than the nutrients are also added to the culture medium to prevent foam formation and to enhance the availability of hydrophobic materials (e.g. lipids) to the cells (Zhang et al., 1992). The following typical mechanisms have been identified which are responsible for cell damage.

(a) Death of those cells which are attached to the rupturing bubbles.
(b) Entrapment of cells in the foam, and
(c) Cell damage by hydrodynamic forces.

Surface active agents, including Pluronic F68, Polyvinyl alcohol, and Methocel, can be added to the culture medium to prevent cell damage by bubble rupturing. These substances rapidly lower air-medium interfacial tension and prevent the attachment of cells with the bubbles. Additionally, hydrophilic polymers, including polyethylene glycol and polyvinyl pyrrolidone, can be added to the medium to prevent the entrapment of cells in the foam, thereby preventing cell damage. Polyethylene glycol and polyvinyl alcohol also protect the cells from hydrodynamic forces (McDowell et al., 1998).

The physical state of the culture medium is an important factor in its storage and transportation during large-scale production. In large quantities, the liquid culture medium can be difficult to transport because of its bulkiness. Additionally, the stability of substances in the medium is generally lower in liquid form than in solid form. On the other hand, the solid culture medium is easy to transport and has high stability. Still, it may present the challenge of non-uniform composition, particularly for substances present in low concentrations, such as Phenol red. The segregation of materials can cause this due to varying particle size distributions. To address these issues, a liquid preconcentrate culture medium is effective at a large scale. In this case, medium components are separated based on their solubility in solvents and then diluted with the appropriate grade of water at the time of use (Gorsky, 2005).

13.3.2 Bioreactor Operation Scale-Up

Stirred tank bioreactors are most widely used for the production of biotechnology-derived products (Fig. 13.1). The physicochemical properties of the culture medium, including temperature, pH, dissolved oxygen concentration, and osmolarity, generally remain the same as those of the bench scale process and considered a scale-independent parameter. On the other hand, mixing rate and hydrodynamic stress to the cells are scale-dependent parameters.

Stirred tank bioreactors are cylindrical vessels with an aspect ratio varying from 1:1 to 3:1, and impellers should be spaced in such a way that there should be no inhomogeneity from top to bottom. However, oxygen transfer is a primary challenge in the large-scale production of biotechnology products which can be avoided by maintaining an aspect ratio of 1. Furthermore, hemispherical bottoms are preferred to avoid the formation of any dead zones in the vessel. Another major challenge in large-scale production is cell damage due to entrapment in foam, which can be minimized by keeping 20–30% of the vessel empty at the top.

Tank geometry also plays a crucial role in the homogeneity of bioreactors and heat and mass transfer rate while mixing as the scale increases. Generally, mixing time substantially increases with an increase in aspect ratio (Garcia-Ochoa & Gomez, 2009).

Another important implication of the increase in bioreactor size is the limited availability of oxygen at a large scale which sometimes may result in a shift towards anabolic metabolism at a large scale (Lara et al., 2009).

13.3.3 Chromatography Process Scale-Up

Chromatography is the most widely used technique in the biotechnology industry or the purification of biomolecules. Purification of products at a large scale needs a scale-up of chromatographic operations, which increases the column volumes. The objective can be accomplished using large columns or running multiple purification

Fig. 13.1 Bioreactor

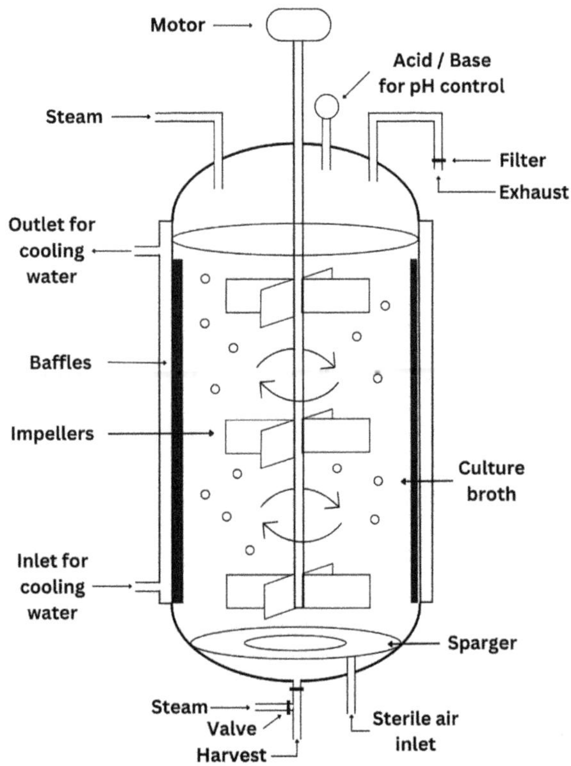

cycles. The following general guidelines can be used for the calculation of operating parameters during the scale-up of the chromatography process (Gorsky, 2005).

(a) Select the column of appropriate diameter.
(b) Calculate the required gel volume using bed height and gel capacity.
(c) Keep the linear velocity constant.
(d) Maintain minimum bed height during large-scale production equal to or greater than 10 cm.

Generally, narrow diameter columns result in an increase in processing time due to the constant linear velocity. On the other hand, in the case of wide columns, processing time decreases due to the increase in flow rate.

Example 13.1 During a chromatography operation, a column of a 60 cm diameter was used for the purification of peptides. The actual column volume was 50 L. Calculate the actual bed height and flow rate if the linear velocity is 300 cm/h. Calculate the flow rate if the column diameter of 100 cm and a bed height of 10 cm were used to scale-up the same chromatographic process.

Actual column volume = 50 L
Column volume $= \pi \times r^2 \times h$

$50 \times 10^3 = 3.14 \times 30 \times 30 \times h$
$h = \frac{50 \times 10^3}{3.14 \times 30 \times 30}$
$h = 17.7$ cm
Calculation of flow rate
Linear velocity = 300 cm/h
Time required to flow = $\frac{17.7}{300}$
Time required to flow = 0.059 h
Time required to flow = 3.54 min
Flow rate = $\frac{50}{3.54}$
Flow rate = 14.1 L/min
Column volume = $\pi \times r^2 \times h$
Column volume = $3.14 \times 50 \times 50 \times 10$
Column volume = 79 L
Time required for complete flow = $\frac{10}{300}$
Time required for complete flow = 2 min
Linear velocity was kept constant
Flow rate = $\frac{\text{Volume}}{\text{Time}}$
Flow rate = $\frac{79}{2}$
Flow rate = 39.5 L/min

13.4 Viral Removal

Viral removal is an essential step in the production of biotechnology-based products, as viruses can potentially contaminate the final product and pose a risk to human health. The viral removal step is typically accomplished using physical, chemical, and/or biological methods.

Physical methods include filtration, which can effectively remove viruses based on their size and physical properties. For example, depth filters, such as glass fiber filters, have been shown to effectively remove viruses with a diameter larger than approximately 0.2 microns. Additionally, membrane filtration, such as ultrafiltration, can also be used to remove viruses based on their size and charge (ElHadidy et al., 2013).

Chemical methods include viral inactivation agents, such as solvent-detergent treatment, which can effectively inactivate viruses by disrupting their structural integrity. Additionally, chemical agents such as formaldehyde and hydrogen peroxide can inactivate viruses by disrupting their genetic material (Elveborg et al., 2022).

Biological methods include the use of viral clearance steps, such as treatment with virus-specific antibodies or viral-specific enzymes. These methods can effectively remove or inactivate viruses by binding to critical viral components (Prusoff et al., 1984).

It is important to note that the choice of viral removal method will depend on the specific product, process, and target virus, and validation of the method is essential. Additionally, the viral removal step should be performed as early as possible to minimize the risk of viral contamination during downstream processing.

13.5 Conclusion

In conclusion, the production of biotechnology-based products is a complex process that requires a thorough understanding of the unit operations involved and the considerations needed to increase batch size. Large-scale production presents unique challenges, such as preventing cell damage and ensuring adequate oxygen supply to the cells. The chapter highlights the importance of oxygen transfer, mixing, temperature, pH, and shear stress in the cell culture process. Additionally, the importance of cleaning, sterilization, and harvesting in maintaining the purity and batch-to-batch reproducibility of biotechnology-based products is emphasized. Advances in cell culture technology, bioreactor processes, and molecular biology techniques have greatly improved the field, and a continued focus on these areas will help to overcome the challenges of large-scale production and meet market demand for biotechnology-based products.

Disclaimer The views expressed in this chapter reflect those of the author and do not necessarily represent the view of their employer.

References

Bareither, R., & Pollard, D. (2011). A review of advanced small-scale parallel bioreactor technology for accelerated process development: Current state and future need. *Biotechnology Progress, 27*(1), 2–14. Available at: https://doi.org/10.1002/btpr.522

Cherry, R. S., & Papoutsakis, E. T. (1986). Hydrodynamic effects on cells in agitated tissue culture reactors. *Bioprocess Engineering, 1*(1), 29–41. Available at: https://doi.org/10.1007/BF00369462

Chisti, Y. (2000). Animal-cell damage in sparged bioreactors. *Trends in Biotechnology, 18*(10), 420–432. Available at: https://doi.org/10.1016/S0167-7799(00)01474-8

Clavijo Rivera, E., et al. (2020). Cross-flow filtration for the recovery of lipids from microalgae aqueous extracts: Membrane selection and performances. *Process Biochemistry, 89*, 199–207. Available at: https://doi.org/10.1016/j.procbio.2019.10.016

Deb, P. K., et al. (2019). Biotechnology-based pharmaceutical products. In *Biomaterials and bionanotechnology* (pp. 153–189). Elsevier. Available at: https://doi.org/10.1016/B978-0-12-814427-5.00005-6

Duong, P., et al. (2019). Cushioned-Density Gradient Ultracentrifugation (C-DGUC) improves the isolation efficiency of extracellular vesicles. *PLoS One, 14*(4), e0215324. Available at: https://doi.org/10.1371/journal.pone.0215324

ElHadidy, A. M., Peldszus, S., & van Dyke, M. I. (2013). An evaluation of virus removal mechanisms by ultrafiltration membranes using MS2 and φX174 bacteriophage. *Separation and Purification Technology, 120*, 215–223. Available at: https://doi.org/10.1016/j.seppur.2013.09.026

Elveborg, S., Monteil, V., & Mirazimi, A. (2022). Methods of inactivation of highly pathogenic viruses for molecular, serology or vaccine development purposes. *Pathogens, 11*(2), 271. Available at: https://doi.org/10.3390/pathogens11020271

Fu, H., et al. (2003). The separation of biomolecules using capillary electrochromatography. *Current Opinion in Biotechnology, 14*(1), 96–100. Available at: https://doi.org/10.1016/S0958-1669(02)00006-X

Garcia-Ochoa, F., & Gomez, E. (2009). Bioreactor scale-up and oxygen transfer rate in microbial processes: An overview. *Biotechnology Advances, 27*(2), 153–176. Available at: https://doi.org/10.1016/j.biotechadv.2008.10.006

Gorsky, I. (2005). Chapter 3: A parenteral drug scale-up. In M. Levin (Ed.), *Pharmaceutical process scale-up* (2nd ed.). CRC Press. Available at: https://doi.org/10.1201/9781420026658

Hu, W., Berdugo, C., & Chalmers, J. J. (2011). The potential of hydrodynamic damage to animal cells of industrial relevance: Current understanding. *Cytotechnology, 63*(5), 445–460. Available at: https://doi.org/10.1007/s10616-011-9368-3

Jildeh, Z. B., Wagner, P. H., & Schöning, M. J. (2021). Sterilization of objects, products, and packaging surfaces and their characterization in different fields of industry: The status in 2020. *Physica Status Solidi, 218*(13), 2000732. Available at: https://doi.org/10.1002/pssa.202000732

Lara, A. R., et al. (2009). Fast dynamic response of the fermentative metabolism of *Escherichia coli* to aerobic and anaerobic glucose pulses. *Biotechnology and Bioengineering, 104*(6), 1153–1161. Available at: https://doi.org/10.1002/bit.22503

Maiorella, B., et al. (1991). Crossflow microfiltration of animal cells. *Biotechnology and Bioengineering, 37*(2), 121–126. Available at: https://doi.org/10.1002/bit.260370205

McDowell, C. L., Carver, R. T., & Papoutsakis, E. T. (1998). Effects of methocel A15LV, polyethylene glycol, and polyvinyl alcohol on CD13 and CD33 receptor surface content and metabolism of HL60 cells cultured in stirred tank bioreactors. *Biotechnology and Bioengineering, 60*(2), 251–258. Available at: https://doi.org/10.1002/(SICI)1097-0290(19981020)60:2<251::AID-BIT12>3.0.CO;2-P

Nienow, A. W. (2006). Reactor engineering in large scale animal cell culture. *Cytotechnology, 50*(1–3), 9–33. Available at: https://doi.org/10.1007/s10616-006-9005-8

Ó'Fágáin, C., Cummins, P. M., & O'Connor, B. F. (2011). Gel-filtration chromatography. *Methods in Molecular Biology, 681*, 25–33. Available at: https://doi.org/10.1007/978-1-60761-913-0_2

Prusoff, W. H., et al. (1984). Overview of the possible targets for viral chemotherapy. In *Targets for the design of antiviral agents* (pp. 1–27). Springer US. Available at: https://doi.org/10.1007/978-1-4684-4709-5_1

Spier, R. E. (2003). In R. E. Spier (Ed.), *Encyclopedia of cell technology*. Wiley. Available at: https://doi.org/10.1002/0471250570

Tanaka, N., et al. (1993). Stationary phase effects in reversed-phase liquid chromatography. *Journal of Chromatography A, 656*(1–2), 265–287. Available at: https://doi.org/10.1016/0021-9673(93)80805-I

Yasin, M., et al. (2019). Bioreactors, gas delivery systems and supporting technologies for microbial synthesis gas conversion process. *Bioresource Technology Reports, 7*, 100207. Available at: https://doi.org/10.1016/j.biteb.2019.100207

Zhang, S., Handa-Corrigan, A., & Spier, R. E. (1992). Foaming and media surfactant effects on the cultivation of animal cells in stirred and sparged bioreactors. *Journal of Biotechnology, 25*(3), 289–306. Available at: https://doi.org/10.1016/0168-1656(92)90162-3

Index

A. B. Jindal (ed.), *Pharmaceutical Process Engineering and Scale-up Principles*, AAPS Introductions in the Pharmaceutical Sciences 13,
https://doi.org/10.1007/978-3-031-31380-6

GPSR Compliance
The European Union's (EU) General Product Safety Regulation (GPSR) is a set of rules that requires consumer products to be safe and our obligations to ensure this.

If you have any concerns about our products, you can contact us on

ProductSafety@springernature.com

In case Publisher is established outside the EU, the EU authorized representative is:

Springer Nature Customer Service Center GmbH
Europaplatz 3
69115 Heidelberg, Germany

www.ingramcontent.com/pod-product-compliance
Ingram Content Group UK Ltd.
Pitfield, Milton Keynes, MK11 3LW, UK
UKHW022211051026
15624UKWH00001B/115
9783031313790